°£16·20
3/82
mc

D1433735

Calculator Programming
for
Chemistry and the Life Sciences

Calculator Programming
for
Chemistry and the Life Sciences

FRANK H. CLARKE

Pharmaceuticals Division
CIBA –GEIGY Corporation
Ardsley, New York

1981

ACADEMIC PRESS
A Subsidiary of Harcourt Brace Jovanovich, Publishers

New York   London   Toronto   Sydney   San Francisco

COPYRIGHT © 1981, BY ACADEMIC PRESS, INC.
ALL RIGHTS RESERVED.
NO PART OF THIS PUBLICATION MAY BE REPRODUCED OR
TRANSMITTED IN ANY FORM OR BY ANY MEANS, ELECTRONIC
OR MECHANICAL, INCLUDING PHOTOCOPY, RECORDING, OR ANY
INFORMATION STORAGE AND RETRIEVAL SYSTEM, WITHOUT
PERMISSION IN WRITING FROM THE PUBLISHER.

ACADEMIC PRESS, INC.
111 Fifth Avenue, New York, New York 10003

United Kingdom Edition published by
ACADEMIC PRESS, INC. (LONDON) LTD.
24/28 Oval Road, London NW1 7DX

Library of Congress Cataloging in Publication Data

Clarke, Frank H.
    Calculator programming for chemistry and the life
sciences.

    Includes bibliographical references and index.
    1. Chemistry, Pharmaceutical--Computer programs.
2. Biological research--Computer programs.
3. Programmable calculators. I. Title.
RS418.C55   542'.8   81-15046
ISBN 0-12-175320-4   AACR2

PRINTED IN THE UNITED STATES OF AMERICA

81 82 83 84   9 8 7 6 5 4 3 2 1

Contents

Preface

This book illustrates the power of the programmable calculator as a tool that provides new dimensions to scientific research. Calculations that once were tedious are now easily performed and provide the scientist with a freedom to explore areas he would not otherwise have considered. It is not possible to be complete in a book of this nature. It is enough if the illustrations with specific and detailed examples encourage others to experiment with such fascinating and challenging problems.

Many people helped to make this book possible. I want to express my appreciation especially to Roland Winter of CIBA–GEIGY Plastics and Additives Division for introducing me to calculator programming in the first place; to Professor Jon Clardy of Cornell University for providing me with Eq. (2-4) for the conversion of x-ray crystallographic coordinates to orthogonal coordinates; to Dennis Artman of CIBA–GEIGY Analytical Research for helpful discussions of potentiometric titrations; to Murray Selwyn of CIBA–GEIGY Pharmaceuticals Division for Eqs. (3-13) and (3-14) for the confidence intervals of the parameters derived by nonlinear regression; to Barry Ritter, formerly of CIBA–GEIGY, who worked with me on early design of the statistics programs, and to John Belanger and Daniel Ben-David who worked with me to perform the potentiometric titrations. I am especially grateful to James Henkel, of the University of Connecticut School of Pharmacy who demonstrated that these programs can be written in Reverse Polish Notation by providing equivalent programs for Chapter 3. My appreciation is also expressed to Mrs. Dorothy Vivian of the Chemistry Division, CIBA–GEIGY Pharmaceuticals, for typing and retyping the manuscript.

Frank H. Clarke

Calculator Programming
for
Chemistry and the Life Sciences

Introduction

The pocket calculator enables chemists and biologists to solve problems that once required computer assistance [1,2]. In fact, the calculator provides approaches to experimental design and data interpretation that otherwise would not be available to the average student or laboratory scientist. It is the purpose of this book to illustrate with specific, detailed examples this new capacity for research. The examples selected fall into three main categories: molecular shapes, potentiometric titrations, and regression analysis. The first and last of these usually involve computers but the calculator enables the scientist to explore their potential on his own. Calculators are now used in acid–base titrations [2], and new methodology has been provided with computers (see references in Chapter 3), which is now available to calculator users as well. A convenient method is provided thereby for the determination of partition coefficients that may aid research in the life sciences.

The programs presented in this book are practical and will be useful for students and scientists with no experience in the use of computers. They are illustrated with specific examples and the instructions are simple and may be used directly. However, the design of the programs is described in detail with the help of decision maps and program notes. Thus, the reader is encouraged to change the programs to suit a particular need or adapt them to meet the requirements of a different calculator.

The programs in this book make full use of the Texas Instruments TI-59 programmable calculator and the PC100A printer attachment. The owner of a Texas Instruments TI-59 programmable calculator [3] or of a Hewlett-Packard HP-41C [4] will find that the corresponding instruction manual fully explains the elements of calculator programming including the use of decision maps. Some of the programs can be used without the printer but most of them are so complicated that the printer is required to use them to their best advantage.

1

Three of the programs of Chapter 3 have been transcribed for use with Hewlett-Packard calculators that use Reverse Polish Notation (RPN). Two of these are for the HP-41C and one is for the HP-67 calculator. The programs of Chapters 2 and 4 use the Master Library Module of the TI-59 calculator to find the determinant and inverse of a matrix. Corresponding programs for the HP-41C calculator may require the Matrix Operations program of the Mathematics Application Pac (or its equivalent).

The programs are written with the laboratory scientist in mind. Data input is often requested by the printer so that the user does not need to remember the order of entry. The printer request becomes a label and when the calculated results are also labeled by the printer, the printout is a record for entry into the laboratory notebook. Many of the programs are complex—one uses 79 data storage memories, most of them several times over during a calculation. The iterative programs may require 30 min or so to reach an answer that is finally printed. In this case, there are built-in safeguards, such as the flashing of interim results that assure the user that a solution to the problem is being approached.

Chapter 1 presents two smaller programs on percentage composition and molecular formula calculations. The former is a simple calculation, but the program allows easy access to the percentage composition of mixtures, which is especially useful for salts and solvates. The second uses carbon, hydrogen, and nitrogen analyses to provide quickly an empirical formula.

A few notes will provide a background for the other programs of the book that are designed to assist in the cooperative interaction of chemistry and biology. Nowhere is this interaction more challenging than in the study of ligand–receptor (or drug–receptor) interactions. An excellent discussion of computer applications in this area is provided by P. Gund, J. B. Rhodes, and G. M. Smith in a recent article entitled, "Three-Dimensional Molecular Modeling and Drug Design" [5]. The authors describe the enormous capacity of the computerized display console to provide the researcher with a three-dimensional picture of molecular shapes and the interactions of bioorganic molecules. Although the calculator is slow and cumbersome compared to the computer, it can be used to advantage in conjunction with a set of molecular model components [6]. The student, working at his own desk, can duplicate much of what the computer display accomplishes, and there is the additional satisfaction of working with physical models. The programs of Chapter 2 illustrate this application with practical examples. The mathematical background is provided for the calculations that are made possible by the capacity of the Master Library Module to perform matrix calculations.

A drug reaches its receptor by crossing biological membranes. This

transport involves a partitioning, sometimes repeatedly, of the drug between aqueous and lipid phases. A recent review entitled, "Lipophilicity
and Drug Activity," by Kubinyi [7] describes and interprets the partition
coefficient, which is used in the study of this phenomenon. Hansch and
Leo have provided a tabulation of partition coefficients [8], and Hansch
described their usefulness in quantitative approaches to pharmacological
structure–activity relationships (QSAR) [9]. Calculator techniques for finding the endpoint of a potentiometric titration are described in Chapter 3.
The reader is then shown how titrations in the presence and absence of
octanol provide a convenient method for determining partition coefficients. A measure of the accuracy of the results is provided by one of the
programs that calculates the titration curve with and without octanol. The
equations involved are derived for the reader and the power of the intera-
tive method to solve nonlinear equations is illustrated.

The partition coefficient is only one of a number of physical properties
that may be correlated with biological activity [9]. The use of regression
analysis in such correlations has been critically reviewed by Martin in
"Quantitative Drug Design" [10]. The study of QSAR involves chemists
and biologists working in close collaboration with computer specialists.
Regression analysis can be used to correlate other phenomena as well
[11]. Programmable calculators are provided with a built-in capacity to
perform simple correlations and a statistics module provides additional
capacity. However, in Chapter 4 the reader is taken much further. Programs are provided for regression analyses involving up to and including
five variables. Regression coefficients are provided together with their
95% confidence limits. The programs calculate the Student's t value. The
correlation coefficient and the F value are also obtained; provision is made
for the addition and subtraction of data points. A special feature is the
provision of converting trivariate data to any of three combinations of
bivariate data. A program is also provided for solving the bilinear equation
of Kubinyi [7].

These programs in no way diminish our dependence on the computer
specialist, but they do provide the scientist with a tool to look critically at
his own data, to study it in various ways, and to have a much better
understanding of how the computer specialist can help.

The reader will appreciate that the programs described in this book are
particular ones used by the author in his own research. No attempt is
made to cover the wide range of practical problems that can be solved
with the calculator. Barnes and Waring in their recent book, "Pocket
Programmable Calculators in Biochemistry" [2], provide a wide variety of
solutions to such problems. They also discuss the Hewlett-Packard HP-
67/97 calculator and the compatibility of the HP-41C calculator, which has

increased capacity. Comparison of the three programs in Chapter 3, which are provided in both algebraic and Reverse Polish Notation, will help the user of the HP-41C calculator to adapt the other programs to his own requirements. Some program steps may be confusing to a Hewlett-Packard user. Sequences such as RCL 01 ÷ (+/− + RCL 02) = which occur in calculating pKa in the programs of Chapter 3 save one program step. They should be changed to RCL 01 ÷ (RCL 02 − RCL 01) = before being transcribed into reverse Polish notation for the Hewlett-Packard calculator. Programs in Chapters 2 and 4, which use the Master Library Module for matrix calculations, conserve program space by inserting pointers in the program rather than calling on the module to do so. The memory locations are selected to correspond to the requirements of the Master Library Module. For these details the TI Programmable 58/59 Master Library Instruction Manual should be consulted.

The book is designed for easy use. Programs are listed together with instructions and each program is illustrated with examples. The mathematical background to the programs is provided. For those interested in program design, there are decision maps to assist in following the programs. Tables of register contents and labels are provided and, in addition to a general description of program design, there are detailed notes accompanying each program. Diagrams and figures are used liberally to help with the explanations. Finally, each chapter (except Chapter 1) has its own list of references to the original literature.

REFERENCES

1. B. Clare, Calculated freedom—how you can dispense with the mainframe computer. *Chem. Br.* **16**, 249, 1980.
2. J. E. Barnes and A. J. Waring, "Pocket Programmable Calculators in Biochemistry." Wiley, New York, 1980.
3. "Personal Programming," TI Programmable 58/59 Owner's Manual, Texas Instruments, Inc., Dallas, Texas, 1977.
4. "Owner's Handbook and Programming Guide, HP-41C," Hewlett-Packard Company, Corvallis, Oregon, 1979.
5. P. Gund, J. D. Androse, J. B. Rhodes, and G. M. Smith, Three-dimensional molecular modeling and drug design. *Science* **208**, 1425, 1980.
6. F. H. Clarke, H. Jaggi, and R. A. Lovell, Conformation of 2,9-dimethyl-3'-hydroxy-5-phenyl-6,7-benzomorphan and its relation to other analgetics and enkephalin. *J. Med. Chem.* **21**, 600, 1978.
7. H. Kubinyi, Lipophilicity and drug activity. *In* "Progress in Drug Research" (E. Jucker, ed.), p. 97. Birkhäuser, Basel, 1979.
8. C. Hansch and A. Leo, "Substituent Constants for Correlation Analysis in Chemistry and Biology." Wiley, New York, 1979.

9. C. Hansch, Quantitative approaches to pharmacological structure–activity relationships. *In* "Structure–Activity Relationships" (C. J. Cavallito, ed.), p. 75. Pergamon, New York, 1973.

10. Y. C. Martin, "Quantitative Drug Design." Marcel Dekker, New York, 1978.

11. N. B. Chapman and J. Shorter, "Correlation Analysis in Chemistry." Plenum, New York, 1978.

CHAPTER 1

Molecular Formulas

I. INTRODUCTION

A. Program 11, Percentage Composition

The programmable calculator offers advantages even for such seemingly routine calculations as those of molecular weight and percentage composition. Program 11 (percentage composition) illustrates this and the great versatility that is achieved with personal programming. The calculations themselves are simple and the program is not complicated. Nevertheless, the results are immediately useful to all organic chemists who need to calculate molecular weights or percentage composition.

Program 11 allows for the inclusion of a second component in any proportion that may be an acid involved in salt formation, a solvent of crystallization, or the second component of a mixture. The steps are shortened when the solvent is water. A third component may easily be added. The program may be modified slightly and used without a printer. However, it was designed for use with a printer and rapidly provides a printout of the formula calculated and the carbon, hydrogen, and nitrogen analyses to two decimal places. The molecular weight is also printed and the input and results are labeled. When water is a solvent of crystallization, the percentage of water is provided.

B. Program 12, Empirical Formula

When the results of the carbon, hydrogen, and nitrogen analyses do not agree with the calculated percentage composition, it is often possible to show that the analysis is probably "off" because of the incomplete removal of a solvent of crystallization. Program 11 is very useful for examining this possibility. However, it may be that the compound analyzed is not what is anticipated or it may be that the empirical formula is not known at all. It is a simple matter of arithmetic to calculate an empirical formula that will give the computed percentage composition (or reasonably close values), and the empirical formula calculated in this way is often useful at least in eliminating unlikely possibilities for the formula of the unknown compound. Although the usefulness of such a calculation depends on the precision of the experimental analysis and the purity of the sample analyzed, program 12 (empirical formula) provides an empirical formula in C, H, N, and sometimes O as well for any set of carbon, hydrogen, and nitrogen analyses. It also provides the exact atomic ratios that assist in determining how precisely the analysis fits the derived formula.

II. CALCULATIONS

A. Program 11, Percentage Composition

For program 11, the percentage of an element present in a compound is provided by Eq (1-1):

$$\% \text{ Element} = \frac{\text{Weight sum of the element}}{\text{Molecular weight}} \times 100 \qquad (1\text{-}1)$$

The weight sum is the atomic weight of the element multiplied by the number of gram-atoms present. For water, the molecular weight of water is multiplied by the number of moles of water and substituted as its weight sum. The atomic weights are those provided in the Merck Index, Ninth Edition, 1976.

B. Program 12, Empirical Formula

For program 12, the atomic ratio of each element in a compound is obtained by dividing the percentage of the element present in the compound by its atomic weight. When these ratios are normalized, the result is the empirical formula. The molecular formula may actually be a multiple of the empirical formula.

Program 12 assumes that carbon, hydrogen, and nitrogen analyses are available. Unless oxygen is the only other element present, only carbon, hydrogen, and nitrogen contributions to the empirical formula are obtained. When oxygen is present, and there are no additional elements, the percentage of oxygen is calculated as the difference between 100% and the sum of the percentages of carbon, hydrogen, and nitrogen. The empirical formula is a multiple of the least-abundant element for each of the other elements, respectively. This is a simple calculation when only carbon, hydrogen, and nitrogen are involved. One simply divides the atomic ratios of carbon and hydrogen by that of nitrogen and rounds off the numbers thus obtained. When oxygen is the fourth element and is least abundant, one is tempted to simply divide each of the other ratios by that of oxygen. This does not give a satisfactory result in most situations because the ratio for oxygen is based on an analysis that was obtained indirectly. A satisfactory result is obtained when all the ratios are first multiplied by a whole number factor representing the ratio of nitrogen to oxygen atoms if the ratio is greater than 1. If not, the reciprocal of this ratio is used as the factor. Program 12 has been designed to carry out the calculation in this manner.

TABLE 1-1

Derivation of Empirical Formula from Percentage Composition

Atom	Percent present	Atomic ratio	Revised ratio(1)	Revised ratio(2)	Rounded ratio
C	68.39	5.6939	17.0818	15.0010	15
H	9.59	9.5139	28.5417	25.0648	25
N	15.95	1.1387	3.4161	3.0000	3
O	6.07	0.3794	1.1382	0.9995	1

The calculations of program 12 are illustrated by the data in Table 1-1 for $C_{15}H_{25}N_3O$. The percent oxygen is the difference between 100% and the sum of the percentages of the other elements. Atomic ratios are tabulated in column 3. These are multiplied by the factor 3 in column 4. In column 5 the ratios of column 4 have been divided by the original atomic ratio for nitrogen, 1.1387. The last column, with rounded numbers, provides the calculated empirical formula based on the analytical data.

III. EXAMPLES

A. Percentage Composition

Examples of percentage compositions are illustrated in Table 1-2, which provides the calculator printouts obtained.

(1) $C_{23}H_{21}N$. The number 23 is put into the display and A is pressed. The entry is printed as 23C. Next, 21 is entered and B is pressed followed by 1 and C. The paper advances when E is pressed to begin the calculation. The final results are printed in quick succession.

(2) $C_{21}H_{14}O$. This is similar to example 1 except that instead of label C, label D is pressed for 1 oxygen atom.

(3) $C_{20}H_{24}N_2 \cdot \frac{1}{2}H_2O$. The parent compound is entered and calculated as in example 1. Then C' is pressed. For one-half mole of water of crystallization, the number 0.50 is entered. E is then pressed and the results are printed as shown. Without the printer, the first number that appears after E is pressed is the percentage of carbon. After the molecular weight, the last number to be displayed is the percentage of water.

(4) $C_{23}H_{32}N_2O \cdot 2HCl \cdot \frac{1}{2} H_2O$. This example is similar to example (3) except that after calculating the percentage composition of the anhydrous hydrochloride salt, C' is pressed and the calculation is repeated with one-half mole of water.

(5) $(C_{10}H_{14}ClNO)_2 \cdot C_2H_2O_4$. In this case, oxalic acid is regarded as the parent compound and the second component is multiplied by the factor of 2.

(6)   Other examples may easily be tested. For instance, in example 4 a second 0.5 mole of water could have been added, or one equivalent of HCl could have been subtracted.

B. Empirical Formula

Three examples of the calculation of empirical formulas are illustrated in Table 1-3. The first example, $C_{15}H_{25}O$ has the same data used to con-

TABLE 1-2

Percentage Compositions

(1) $C_{23}H_{21}N$		(4) $C_{23}H_{32}N_2O$		(5) $C_2H_2O_4$	
23.	C	23.	C	2.	C
21.	H	32.	H	2.	H
1.	N	2.	N	4.	O
		1.	O		
88.71	%C			26.68	%C
6.80	%H	78.37	%C	2.24	%H
4.50	%N	9.15	%H	0.00	%N
311.43	MW	7.95	%N	90.04	MW
		352.52	MW		

(2) $C_{21}H_{14}O$

$C_{23}H_{32}N_2O \cdot 2HCl$

$C_2H_2O_4 \cdot 2C_{10}H_{14}ClNO$

21.	C			10.	C
14.	H			14.	H
1.	O	1.	H	1.	CL
		1.	CL	1.	N
89.34	%C	2.00	M	1.	O
5.00	%H	64.93	%C	2.00	M
0.00	%N	8.05	%H	53.99	%C
282.34	MW	6.58	%N	6.18	%H
		425.44	MW	5.72	%N
				489.39	MW

(3) $C_{20}H_{24}N_2$

$C_{23}H_{32}N_2O \cdot 2HCl \cdot \frac{1}{2}H_2O$

20.	C		
24.	H		H2O
2.	N	0.50	M
		63.59	%C
82.15	%C	8.12	%H
8.27	%H	6.45	%N
9.58	%N	434.45	MW
292.42	MW	2.07	%X

$C_{20}H_{24}N_2 \cdot \frac{1}{2}H_2O$

	H2O
0.50	M
79.69	%C
8.36	%H
9.29	%N
301.43	MW
2.99	%X

struct Table 1-1. The raw data are theoretical values obtained from program 11. Note that rounding of the atomic ratios is not really necessary when the analyses are accurate. However, if the analyses have the normal error (i.e., ± 0.2%), there can be considerable doubt about the correct empirical formula. In each example of Table 1-3, the theoretical raw data are presented first and then a typical example of experimental data. Note

TABLE 1-3

Empirical Formulas

(1) $C_{15}H_{25}N_3O$			(2) $C_{14}H_{23}NO_2$			(3) $C_{20}H_{32}ClNO_2S$		
FOUND			FOUND			FOUND		
	68.39	%C		70.85	%C		62.23	%C
	9.59	%H		9.77	%H		8.36	%H
	15.95	%N		5.90	%N		3.63	%N
	6.07	%O		13.48	%O	CALCD		
CALCD			CALCD				19.9921	RC
	15.0010	RC		14.0041	RC		32.0025	RH
	25.0648	RH		23.0106	RH		1.0000	RN
	3.0000	RN		1.0000	RN			
	0.9995	RO		2.0003	RO		20.	C
							32.	H
	15.	C		14.	C		1.	N
	25.	H		23.	H			
	3.	N		1.	N			
	1.	O		2.	O			

						FOUND		
							62.07	%C
							8.14	%H
FOUND			FOUND				3.45	%N
	68.20	%C		70.78	%C	CALCD		
	9.65	%H		9.86	%H		20.9811	RC
	15.80	%N		6.00	%N		32.7861	RH
	6.35	%O		13.36	%O		1.0000	RN
CALCD			CALCD					
	15.1013	RC		13.7570	RC		21.	C
	25.4610	RH		22.8355	RH		33.	H
	3.0000	RN		1.0000	RN		1.	N
	1.0556	RO		1.9494	RO			
	15.	C		14.	C			
	25.	H		23.	H			
	3.	N		1.	N			
	1.	O		2.	O			

that the number of hydrogen atoms could easily be estimated as 26 instead of 25 on the basis of the experimental data for $C_{15}H_{25}N_3O$.

In example 2, $C_{14}H_{23}O_2$, the ratio of nitrogen is now less than that of oxygen so the nitrogen atomic ratio is divided into each of the other atomic ratios to supply the revised ratios. In this example, the experimental error still provides a correct estimate of the number of carbon and hydrogen atoms in the rounded numbers, although neither the carbon or the hydrogen results are very precise in the experimental case.

The last example contains elements other than oxygen in addition to carbon, hydrogen, and nitrogen. For this reason, the empirical formula is only a partial one. For the experimental case, the results are not as close as might have been anticipated.

IV. INSTRUCTIONS

A. Program 11, Percentage Composition

Step 1. *Insert programs 111, 112.* The program is designed for use with a printer. If the printer is not used, Op 06 is replaced by R/S, NOP at 234, 235; 252, 253; 270, 271, 280, 281; and 309, 310.

Step 2. *Initialize.* Press E'. 0 is displayed but not printed.

Step 3. *Enter empirical formula* of parent component by placing the number of atoms in the display and pressing the appropriate label. It does not matter what order is used for this procedure. If a particular atom is not present, the label is not pressed. To enter the number of carbon atoms, press A; for hydrogen atoms, press B; for nitrogen, press C; for oxygen, press D; for bromine, press A'; for chlorine, press B'; and for sulfur, press SBR EE.

Step 4. *Calculate* percentage composition and molecular weight of parent compound: Press E. If the printer is attached, all the results are printed and labeled: %C, %H, %N, MW (for molecular weight). If the printer is not used, the percent carbon is displayed when E is pressed. Pressing R/S repeatedly then provides %H, %N, and molecular weight, respectively, each in turn.

Step 5. *Initialize for second component.* Press D', unless the second component is water, in which case, press C'. Zero is displayed after D' and 230332 after C'. If the printer is used, H_2O is printed when C' is pressed.

Step 6. *Enter empirical formula of second component* (if it is not water; if the second component is water, go directly to step 7). The elements of the second component are entered exactly as in step 3.

Step 7. *Enter number of moles of second component.* A fractional (expressed as a decimal) number may be entered. If a fraction is entered twice, the second value is summed on the first value. Thus, after 0.5 moles of water, the next entry could be 0.5 moles again to provide 1.0 moles of water, or −0.15 moles of water the second time, to give 0.35 moles of water added to the parent compound. If the new entry is the same compound, D' (or C') is not pressed a second time.

Step 8. *Calculate.* Press E. The results are obtained as in step 4 except that the results are for the parent compound plus the additional component.

Step 9. *Repeat as required.* Steps 5, 6, 7, and 8 may be repeated as often as required.

Step 10. *To replace Br with any other element X.* Press E' to initialize, enter the atomic weight of X and store it in R_{14}; then proceed with steps 3–5 as before but now using A' for X instead of for Br.

B. Program 12, Empirical Formula

Step 1. *Insert programs 121, 122.* The program is designed to use with a printer. If the printer is not used, Op 06 is replaced by R/S Nop at 078, 079; 139, 140; 154, 155; 169, 170; 185, 186; 193, 194; 201, 202; and 218, 219.

Step 2. *Initialize.* Press E'. The word FOUND is printed and 12.01 appears in the display as a reminder to enter first the percentage of carbon.

Step 3. *Enter Data.* Percentages of carbon, hydrogen, and nitrogen are entered in turn with R/S pressed after each entry. If it is known that oxygen is the only other element present, R/S is pressed again (or label D). If this is not the case, or if no other elements are present, label E is pressed. In either case, the results are calculated directly and printed with appropriate labels (see Table 1-3). If the printer is not used, each calculated result is displayed in turn as R/S is pressed.

V. DESIGN

A. Program 11, Registers and Flags

Program 11 (percentage composition) operates by dividing the weight sum of the element (C, H, N, or water) by the molecular weight and multiplying the result by 100. The weight sum of the element is the total present in all components.

The storage registers are divided into three banks, A, B, and C (Table 1-4). Bank A contains the atomic weights of the elements and is filled when label E' is pressed. Bank C contains the weight sums of the elements of the last component and of water. These are computed as each element is entered. Bank B contains the weight sums of the elements of the entire molecule. These are summed for each element when E is pressed, before the final percentage is calculated.

Two flags are used: Flag 1 is used to calculate the percentage of water if C' has been pressed. Flag 2 is used to distinguish whether the parent compound is being calculated or not. For the parent compound, the fractional number is automatically 1. For a second component, the fractional number is that entered in the display before E is pressed.

TABLE 1-4

Program 11--Percentage Composition

Register contents		
Elements	Molecule	Component
R10 At. wt. C	R30 Wt sum C	R50 Wt sum C
11 At. wt. H	31 Wt sum H	51 Wt sum H
12 At. wt. N	32 Wt sum N	52 Wt sum N
13 At. wt. O	33 MW	53 MW
14 At. wt. Br	—	54 At. wt. H_2O
15 At. wt. Cl	R44 Fractional number	55 Wt. sum H_2O
17 At. wt. S		
	User defined labels	
A Carbon	A' Bromine	SBR EE Sulfur
B Hydrogen	B' Chlorine	
C Nitrogen	C' H_2O	
D Oxygen	D' Initialize (2)	
E Calculate	E' Initialize (1)	

B. Program 11, Labels

It may be helpful to follow what action the program takes as each label is pressed (see Instructions).

E'—Initialization: Clears memory registers, enters atomic weights into bank A and resets the flags (RST).

A, B, C, D, A', B', or SBR EE. Enters empirical formula. The number of atoms is multiplied by the respective atomic weight and the result is entered into the appropriate register of bank C.

E—Calculate: The first time E is pressed, flag 2 is set. The contents of bank C are transferred to bank B, and C, H, and N percentages are calculated and displayed (or printed). The molecular weight is also displayed (or printed).

C'—Enters atomic weight of hydrogens and the molecular weight of water into bank C. It also sets flag 1.

D'—Clears the memories of bank C in preparation for entry of elements of the second component.

E—Calculate for the second time: When E is pressed the second time, the contents of bank C are multiplied by the fractional number of moles of the second component and summed into bank B before new percentages of C, H, and N are calculated.

The program described above is easy to operate and may be readily changed either to include additional elements or to change the meanings of the labels. The labels may be changed by placing the correct atomic weights in proper position beginning at step 375 and by altering the printing code before Op 04 at the appropriate place in the program. To create a new label, simply key in the new steps at the appropriate position, i.e., at 433 in the present program and include the new label as well in the initialization sequence, i.e., at 398 in the present program. By changing the partitioning and rearranging the register banks more efficiently, many new labels can be incorporated.

C. Program 12, Registers and Flags

Program 12 (empirical formula) uses only 10 registers. These are listed in Table 1-5. Flag 1 is used to distinguish the case wherein O is the only additional element besides C, H, and N.

D. Program 12, Labels

User-defined labels for program 12 are defined in Table 1-5. Initialization by pressing E' places the atomic weights of C, H, and N in the registers, clears all the other memories and resets the flag. The user-defined labels A, B, C, and D can be used in any order to enter the percentages, but if entered in the order C, H, and N, only R/S needs to be

TABLE 1-5

Program 12—Empirical Formula

Register contents	User defined labels
00 Lowest ratio	A Carbon
01 Sum percent	B Hydrogen
02 %C → ratio C	C Nitrogen
03 %H → ratio H	D Oxygen
04 %N → ratio N	E Calculate
05 %O → ratio O	A' Bromine
06 At. wt. C	B' Chlorine
07 At. wt. H	C' H_2O
08 At. wt. N	D' Initialize (component 2)
09 At. wt. O	E' Initialize (component 1)

pressed. One more press of R/S indicates that oxygen is the only additional element, otherwise label E is pressed. As each percentage is entered, the atomic weight of the next element is displayed as a reminder of the order in which the data are most conveniently entered. The whole number multiplier, representing the ratio of nitrogen to oxygen atoms, is obtained by the sequence following LBL IFF. The sequence FIX OO EE INV EE provides the rounded whole number that is not used unless it is greater or equal to 1. Of course, multiplying by 1 does not change anything, but the steps are short so there is no significant time lost.

The program was designed for a printer but can easily be changed for use without a printer as described in the instructions.

VI. PROGRAMS

A. Program Listings

The programs, Percentage Composition (program 11) and Empirical Formula (program 12) can be recorded on two cards if both sides of each card are used. No special partitioning is required. Program steps are listed in Tables 1-6 to 1-9, inclusive, and label locations are listed at the conclusion of each program. Marginal numbers in parentheses refer to explanatory notes that are presented in Section B. The notes help to explain the detailed operation of the programs and are especially useful when program changes are contemplated. Each program step is numbered (first column). This is followed by the two-digit numerical key code (second column) and the key designation (third column). This general format is used in all program listings throughout the book.

The programs are listed in tables, each of which contains a maximum of 240 steps (including step 000) and can be recorded on one side of a card. Program numbers have three digits. The first digit identifies the type of program and the second digit is a more specific designation. The third digit refers to the table in which the program is listed or to the card side on which the program is recorded.

It is important, in entering these programs, not to have a single error. In most cases, even an error in one digit of a key code will make the program inoperative. Programs, once entered, may be listed with the printer and checked against the table. Each program is illustrated by at least one example so that proper program operation may be checked by duplicating the printout of the example.

TABLE 1-6

Program 111 Listing

000	91	R/S		060	44	44		120	43	RCL		180	43	RCL	
001	76	LBL	C	061	69	OP		121	44	44		181	44	44	
002	11	A		062	06	06 print		122	69	OP		182	65	×	
003	42	STO	(1)	063	91	R/S		123	06	06 print		183	43	RCL	
004	44	44		064	76	LBL	O	124	91	R/S		184	51	51	
005	65	×		065	14	D		125	76	LBL		185	95	=	
006	43	RCL		066	42	STO		126	18	C'	H_2O	186	44	SUM	
007	10	10		067	44	44 (1)		127	00	0	(3)	187	31	31	
008	95	= Wt.		068	65	×		128	42	STO		188	43	RCL	
009	44	SUM Sum		069	43	RCL		129	50	50		189	44	44	
010	50	50 C		070	13	13		130	42	STO		190	65	×	
011	44	SUM		071	95	=		131	52	52		191	43	RCL	
012	53	53		072	44	SUM		132	43	RCL		192	52	52	
013	01	1 C		073	53	53		133	11	11		193	95	=	
014	05	5		074	03	3 O		134	65	×		194	44	SUM	
015	69	OP		075	02	2		135	02	2		195	32	32	
016	04	04		076	69	OP		136	95	=		196	43	RCL	
017	43	RCL		077	04	04		137	42	STO		197	44	44	
018	44	44		078	43	RCL		138	51	51		198	65	×	
019	69	OP		079	44	44		139	42	STO		199	43	RCL	
020	06	06 print		080	69	OP		140	53	53		200	53	53	
021	91	R/S		081	06	06 print		141	42	STO		201	95	=	
022	76	LBL	H	082	91	R/S		142	54	54		202	44	SUM	
023	12	B		083	76	LBL	Br	143	43	RCL		203	33	33	
024	42	STO	(1)	084	16	A'		144	13	13		204	43	RCL	
025	44	44		085	42	STO		145	44	SUM		205	44	44	
026	65	×		086	44	44 (1)		146	53	53		206	65	×	
027	43	RCL		087	65	×		147	44	SUM		207	43	RCL	
028	11	11		088	43	RCL		148	54	54		208	54	54	
029	95	=		089	14	14 (2)		149	02	2 H		209	95	=	
030	44	SUM		090	95	=		150	03	3		210	44	SUM	
031	51	51		091	44	SUM		151	00	0 2		211	55	55	
032	44	SUM		092	53	53		152	03	3		212	87	IFF	
033	53	53		093	01	1		153	03	3 O		213	02	02 (9)	
034	02	2 H		094	04	4		154	02	2		214	33	X²	
035	03	3		095	03	3		155	69	OP		215	98	ADV	
036	69	OP		096	05	5		156	04	04		216	76	LBL	
037	04	04		097	69	OP		157	69	OP		217	34	ΓX (10)	
038	43	RCL		098	04	04		158	05	05 print		218	06	6	
039	44	44		099	43	RCL		159	86	STF (4)		219	01	1 %	
040	69	OP		100	44	44		160	01	01		220	01	1 C	
041	06	06 print		101	69	OP		161	91	R/S		221	05	5	
042	91	R/S		102	06	06 print		162	76	LBL		222	69	OP	
043	76	LBL	N	103	91	R/S		163	15	E calc.		223	04	04	
044	13	C		104	76	LBL		164	58	FIX		224	43	RCL	
045	42	STO	(1)	105	17	B'	Cl	165	02	02 (5)		225	30	30	
046	44	44		106	42	STO		166	87	IFF		226	55	÷	
047	65	×		107	44	44 (1)		167	02	02 (6)		227	43	RCL	
048	43	RCL		108	65	×		168	45	Y×		228	33	33	
049	12	12		109	43	RCL		169	01	1 (7)		229	65	×	
050	95	=		110	15	15		170	76	LBL		230	01	1	
051	44	SUM		111	95	=		171	45	Y× (6)		231	00	0	
052	52	52		112	44	SUM		172	42	STO Fract. No.		232	00	0	
053	44	SUM		113	53	53		173	44	44		233	95	=	
054	53	53		114	01	1		174	65	× (8)		234	69	OP	
055	03	3 N		115	05	5		175	43	RCL		235	06	06 print	
056	01	1		116	02	2		176	50	50		236	06	6	
057	69	OP		117	07	7		177	95	=		237	01	1 %	
058	04	04		118	69	OP		178	44	SUM		238	02	2 H	
059	43	RCL		119	04	04		179	30	30		239	03	3	

17

TABLE 1-7

Program 112 Listing

240	69	OP	
241	04	04	
242	43	RCL	
243	31	31	
244	55	÷	
245	43	RCL	
246	33	33	
247	65	×	
248	01	1	
249	00	0	
250	00	0	
251	95	=	
252	69	OP	
253	06	06	print
254	06	6	%
255	01	1	
256	03	3	N
257	01	1	
258	69	OP	
259	04	04	
260	43	RCL	
261	32	32	
262	55	÷	
263	43	RCL	
264	33	33	
265	65	×	
266	01	1	
267	00	0	
268	00	0	
269	95	=	
270	69	OP	
271	06	06	print
272	03	3	M
273	00	0	
274	04	4	W
275	03	3	
276	69	OP	
277	04	04	
278	43	RCL	
279	33	33	
280	69	OP	
281	06	06	print
282	86	STF	(11)
283	02	02	
284	87	IFF	
285	01	01	(4)
286	88	DMS	
287	98	ADV	
288	98	ADV	
289	98	ADV	
290	91	R/S	
291	76	LBL	
292	88	DMS	(4)
293	06	6	%
294	01	1	
295	04	4	X (12)
296	04	4	
297	69	OP	
298	04	04	
299	43	RCL	

300	55	55	
301	55	÷	
302	43	RCL	
303	33	33	
304	65	×	
305	01	1	
306	00	0	
307	00	0	
308	95	=	
309	69	OP	
310	06	06	print
311	98	ADV	
312	98	ADV	
313	98	ADV	
314	91	R/S	
315	76	LBL	
316	19	D'	(13)
317	22	INV	(14)
318	58	FIX	
319	00	0	
320	42	STO	
321	50	50	
322	42	STO	
323	51	51	
324	42	STO	
325	52	52	
326	42	STO	
327	53	53	
328	42	STO	
329	54	54	
330	42	STO	
331	55	55	
332	22	INV	
333	86	STF	(15)
334	01	01	
335	61	GTO	
336	00	00	
337	00	00	
338	76	LBL	
339	10	E'	(16)
340	47	CMS	
341	01	1	
342	02	2	
343	93	.	
344	00	0	
345	01	1	
346	01	1	
347	42	STO	
348	10	10	
349	01	1	
350	93	.	
351	00	0	
352	00	0	
353	07	7	
354	09	9	
355	42	STO	
356	11	11	
357	01	1	
358	04	4	
359	93	.	

360	00	0	
361	00	0	
362	06	6	
363	07	7	
364	42	STO	
365	12	12	
366	01	1	
367	05	5	
368	93	.	
369	09	9	
370	09	9	
371	09	9	
372	04	4	
373	42	STO	
374	13	13	
375	07	7	
376	09	9	
377	93	.	
378	09	9	
379	00	0	
380	04	4	
381	42	STO	
382	14	14	
383	03	3	
384	05	5	
385	93	.	
386	04	4	
387	05	5	
388	03	3	
389	42	STO	
390	15	15	
391	03	3	
392	02	2	
393	93	.	
394	00	0	
395	06	6	
396	42	STO	
397	17	17	
398	22	INV	
399	58	FIX	
400	00	0	
401	81	RST	
402	76	LBL	
403	52	EE	(17)
404	42	STO	
405	44	44	
406	65	×	
407	43	RCL	
408	17	17	
409	95	=	
410	44	SUM	
411	53	53	
412	03	3	
413	06	6	
414	69	OP	
415	04	04	
416	43	RCL	
417	44	44	
418	69	OP	
419	06	06	

420	91	R/S	
421	76	LBL	
422	33	X^2	(18)
423	03	3	M
424	00	0	
425	69	OP	
426	04	04	
427	43	RCL	
428	44	44	
429	69	OP	
430	06	06	print
431	61	GTO	
432	34	ΓX	
433	00	0	
434	00	0	

LABELS

002	11	A
023	12	B
044	13	C
065	14	D
084	16	A'
105	17	B'
126	18	C'
163	15	E
171	45	Y^X
217	34	ΓX
292	88	DMS
316	19	D'
339	10	E'
403	52	EE
422	33	X^2

TABLE 1-8
Program 121 Listing

000	91	R/S		060	43	RCL		120	00	00		180	05	5 C
001	76	LBL		061	01	01 (3)		121	87	IFF		181	69	OP
002	11	A C		062	94	+/-		122	01	01 (10)		182	04	04
003	42	STO		063	85	+		123	87	IFF		183	43	RCL
004	02	02		064	01	1		124	76	LBL (11)		184	02	02
005	44	SUM		065	00	0		125	88	DMS		185	69	OP
006	01	01		066	00	0		126	43	RCL		186	06	06 print
007	06	6 %		067	95	=		127	00	00		187	02	2 H
008	01	1		068	42	STO		128	22	INV		188	03	3
009	01	1 C		069	05	05		129	49	PRD		189	69	OP
010	05	5		070	06	6		130	02	02		190	04	04
011	69	OP		071	01	1		131	03	3 R		191	43	RCL
012	04	04		072	03	3		132	05	5		192	03	03
013	43	RCL		073	02	2		133	01	1 C		193	69	OP
014	02	02 (1)		074	69	OP		134	05	5		194	06	06 print
015	69	OP		075	04	04		135	69	OP		195	03	3 N
016	06	06 print		076	43	RCL		136	04	04		196	01	1
017	43	RCL		077	05	05		137	43	RCL		197	69	OP
018	07	07 (2)		078	69	OP		138	02	02		198	04	04
019	91	R/S		079	06	06 print		139	69	OP		199	43	RCL
020	76	LBL		080	86	STF		140	06	06 print		200	04	04
021	12	B		081	01	01 (4)		141	43	RCL		201	69	OP
022	42	STO		082	76	LBL (5)		142	00	00		202	06	06 print
023	03	03		083	15	E		143	22	INV		203	87	IFF (14)
024	44	SUM		084	58	FIX		144	49	PRD		204	01	01
025	01	01		085	04	04 (6)		145	03	03		205	78	Σ+
026	06	6		086	69	OP		146	03	3 R		206	98	ADV
027	01	1		087	00	00 (7)		147	05	5		207	98	ADV
028	02	2		088	01	1 C		148	02	2 H		208	98	ADV
029	03	3		089	05	5		149	03	3		209	91	R/S
030	69	OP		090	01	1 A		150	69	OP		210	76	LBL (14)
031	04	04		091	03	3		151	04	04		211	78	Σ+
032	43	RCL		092	02	2 L		152	43	RCL		212	03	3
033	03	03		093	07	7		153	03	03		213	02	2
034	69	OP		094	01	1 C		154	69	OP		214	69	OP
035	06	06 print		095	05	5		155	06	06 print		215	04	04
036	43	RCL		096	01	1 D		156	43	RCL		216	43	RCL
037	08	08		097	06	6		157	00	00		217	05	05
038	91	R/S		098	69	OP		158	22	INV		218	69	OP
039	76	LBL N		099	01	01		159	49	PRD		219	06	06
040	13	C		100	69	OP		160	04	04		220	98	ADV
041	42	STO		101	05	05 print		161	03	3 R		221	98	ADV
042	04	04		102	43	RCL		162	05	5		222	98	ADV
043	44	SUM		103	06	06 (8)		163	03	3 N		223	91	R/S
044	01	01		104	22	INV		164	01	1		224	76	LBL
045	06	6		105	49	PRD		165	69	OP		225	89	π (12)
046	01	1		106	02	02		166	04	04		226	43	RCL
047	03	3		107	43	RCL		167	43	RCL		227	00	00
048	01	1		108	07	07		168	04	04		228	22	INV
049	69	OP		109	22	INV		169	69	OP		229	49	PRD
050	04	04		110	49	PRD		170	06	06 print		230	05	05
051	43	RCL		111	03	03		171	87	IFF (12)		231	03	3
052	04	04		112	43	RCL		172	01	01		232	05	5
053	69	OP		113	08	08		173	89	π		233	03	3
054	06	06 print		114	22	INV		174	98	ADV		234	02	2
055	43	RCL		115	49	PRD		175	76	LBL		235	69	OP
056	09	09		116	04	04		176	77	GE (13)		236	04	04
057	91	R/S		117	43	RCL		177	58	FIX		237	43	RCL
058	76	LBL		118	04	04 (9)		178	00	00		238	05	05
059	14	D		119	42	STO		179	01	1		239	69	OP

19

TABLE 1-9

Program 122 Listing

240	06	06		300	01	1
241	98	ADV		301	02	2
242	71	SBR		302	93	.
243	77	GE		303	00	0
244	76	LBL		304	01	1
245	87	IFF	(15)	305	01	1
246	43	RCL		306	42	STO
247	09	09		307	06	06
248	22	INV		308	01	1
249	49	PRD		309	93	.
250	05	05		310	00	0
251	43	RCL		311	00	0
252	04	04		312	08	8
253	55	÷		313	42	STO
254	43	RCL		314	07	07
255	05	05		315	01	1
256	95	=		316	04	4
257	58	FIX	(16)	317	93	.
258	00	00		318	00	0
259	52	EE		319	00	0
260	22	INV		320	07	7
261	52	EE		321	42	STO
262	58	FIX		322	08	08
263	04	04		323	01	1
264	22	INV		324	05	5
265	77	GE		325	93	.
266	88	DMS		326	09	9
267	49	PRD		327	09	9
268	05	05		328	09	9
269	49	PRD		329	42	STO
270	04	04		330	09	09
271	49	PRD		331	43	RCL
272	03	03		332	06	06
273	49	PRD		333	81	RST
274	02	02		334	00	0
275	71	SBR		335	00	0
276	88	DMS		336	00	0
277	76	LBL				
278	10	E'	(17)			
279	01	1				
280	32	X:T		**LABELS**		
281	47	CMS				
282	58	FIX		002	11	A
283	02	02		021	12	B
284	69	OP		040	13	C
285	00	00		059	14	D
286	02	2	F	083	15	E
287	01	1		125	88	DMS
288	03	3	O	176	77	GE
289	02	2		211	78	Σ+
290	04	4	U	225	89	π
291	01	1		245	87	IFF
292	03	3	N	278	10	E'
293	01	1				
294	01	1	D			
295	06	6				
296	69	OP				
297	01	01				
298	69	OP				
299	05	05	print			

B.   Program Notes

1.   Programs 111, 112

(1)   R44 is used initially to store data input so that it may be recalled and printed with a label when the sums have been entered.

(2)   R14 contains the atomic weight of bromine. As explained in instruction 10, this register may be used for that of different element X and A′ used as the label for X.

(3)   Label C′ is an initialization step, which first clears out the weight sums of C and N from bank C and then enters the weight sums of hydrogen and water in the appropriate registers.

(4)   Flag 1 is set so that the percentage of water will be included in the calculations (label DMS, step 292).

(5)   Two places of decimals are usually all that are required. A different setting may be used if desired.

(6)   When flag 2 is set (step 283) the weight sums of the elements of the component in bank C will be multiplied by the fraction entered in the display (label y^x, step 171).

(7)   When the percentages of the first component are calculated, the program puts the number 1 into the display.

(8)   This sequence multiplies the weight sums of bank C and sums them into the corresponding registers of bank B.

(9)   If flag 2 is set, the elements of the second component are being used. In order that the fractional number of moles may be recorded, program operations moves to label X^2 to print and label the fraction (label X^2, step 423). Operation returns to $\sqrt{X}$ at step 217. If flag 2 is not set, it is component 1 that is involved in the calculations and the paper is advanced before its percentages are calculated and printed.

(10)   Label $\sqrt{X}$ begins the calculation of the percentages of C, H, and N and prints the results with appropriate labels.

(11)   Flag 2 is set so that if further data are entered it will be regarded as a second component (see note 6).

(12)   It would be preferable if H_2O were the label here, but the program cannot accommodate the required number of label spaces.

(13)   Label D′ prepares for data entry for a second component.

(14)   Inert fix is used so that whole numbers will appear in the printout without .00, which would appear in fix 2 format.

(15)   Use of label D′ may follow that of label C′, which left flag 1 set (see note 4).

(16)   Label E′ clears the memories, places the atomic weights in memory bank A, inverts the fix decimal format, puts 0 in the display, and with RST resets the flags and sends program operation to the beginning.

(17) Label EE is employed as a user-defined label for sulfur.

(18) Label X^2 prints the fraction with its label (M) for the number of moles of the second component.

2. *Programs 121, 122*

(1) Register 2 is recalled after the printing register has been set up with Op 04.

(2) R07 contains the atomic weight of hydrogen. Its entry into the display reminds the user that the next entry is to be the % H. The R/S key may be used to continue program operation. If the next entry is not the element whose atomic weight is displayed, user-defined labels (A, B, C, D) may be used or O entered and the R/S key pressed again to display the atomic weight of nitrogen.

(3) Label D is pressed only if oxygen is the only other element present. Its percentage is calculated by subtracting the sum of the percentages of C, H, and N from 100.

(4) When label D is pressed, flag 1 is set so that ratio O will be printed (see step 203 and label $\Sigma+$, step 211), (see also note 10).

(5) Label E begins the calculation.

(6) The ratios will be printed to four places of decimals.

(7) Op 00 clears the print memories that have entries from Op 4.

(8) This sequence provides atomic ratios.

(9) The atomic ratio of nitrogen is stored in R00 for later comparison with that of oxygen (step 245).

(10) If flag 1 is set, oxygen is the only other element present and its ratio is compared with that of nitrogen. The lowest ratio is then used to calculate the empirical formula (label IFF, step 245).

(11) This sequence prints and labels the atomic ratios of C, H and N.

(12) If flag 1 is set, label π (step 225) prints the ratio of oxygen.

(13) Label GE prints and labels the rounded atomic ratios as the derived empirical formula.

(14) If flag 1 is set, label $\Sigma+$ (step 211) prints the rounded atomic ratio of oxygen.

(15) Label IFF compares the atomic ratio of nitrogen with that of oxygen to find the lowest one.

(16) This sequence converts the fraction to a rounded whole number. If the fraction is less than 0.5, the whole number is 0. The use of INV GE then sends program operation to label DMS (step 125) where the atomic ratios are printed. On the other hand, if the whole number ratio is 1 or greater (label E′ placed 1 in the t register, note 17), the ratios are multiplied by this new whole number.

(17) Label E′ places 1 in the t register (see note 16), clears the memories, fixes the decimal point at 2, clears the print registers, prints the word FOUND, stores the atomic weights of C, H, N, and O, recalls the atomic weight of C to the display and finally, with RST resets the flags and sends program operation to the beginning.

CHAPTER 2

Coordinate Transformations

I. INTRODUCTION

A. Comparing Molecules

The most effective method of studying the relationships and interactions of bioorganic molecules is to visualize the positions of their component atoms and the bonds between them. Elaborate and expensive computer-based equipment has been designed for this purpose [1]. However, skeletal and space filling models are very useful and much more easily accessible [2–5a,5b]. In this chapter we shall show that x-ray crystallographic data may be used to compare conformations [6a] and to build skeletal molecular models. Tollenaere, Moereels, and Raymaekers have provided a useful monograph in which conformations of drugs are compared [6b].

Models of rigid molecules are readily constructed from model building components [3]. For flexible molecules where many conformations are possible, x-ray coordinate data provide at least one reasonable conformation that is useful for many purposes [4]. Enzymes are extreme examples of such molecules [2] but there are many smaller, flexible molecules for which x-ray data may be used to compare conformations. The use of such data is often left to those with access to a computer. However, it is now practical to make useful comparisons and to build molecular models with coordinate transformations that can be performed easily with a programmable calculator.

The problem of comparing two compounds is not necessarily a trivial one even with models in hand. When two molecular structures are very similar, it is relatively easy to observe the differences between them. However, when two organic compounds have very different structures and both are known to bind to the same receptor, it is often a very difficult problem to measure the degree of correspondence without resorting to complex computer equipment [7–9]. An approach to the problem, which is very effective for most purposes, is not to try to line up all the comparable atoms at once but rather to choose three of the most important atoms or features in each molecule. One of these, B, is placed at the origin of an orthogonal coordinate system and another, A, is placed along the X axis.

The third atom or feature, C, is then placed in a particular orientation relative to the $X-Y$ plane. The molecule is then in what will be referred to as its *standard orientation* (Fig. 2-1c). The comparable features of the second molecule are similarly oriented. With each molecule in such a defined orientation, it is relatively easy to observe the degree of correspondence of all the other features of the two compounds (Fig. 2-1d).

As a simple illustration, consider the molecules of hydrogen sulfide and water. Fig. 2-1a is a photograph of the Corey–Pauling–Kolton (CPK) models of these two compounds. Since these are space filling models made to scale, it is easy to see that the sulfur atom is larger than the oxygen atom. It is difficult, however, to determine whether the bond angles in the

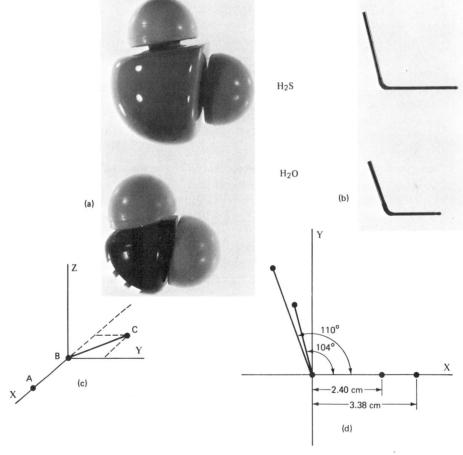

Fig. 2-1   Comparing molecules of H_2S and H_2O: (a) CPK space-filling models; (b) Dreiding skeletal models; (c) molecule ABC placed in a standard orientation; (d) drawing to the scale of 0.4 Å/cm, H–S bond is longer, H–S–H angle is smaller.

two compounds are the same or different. Such measurements are simplified with Dreiding skeletal models, Fig. 2-1b. In these models the lengths of the rods or tubes are proportional to the distances between the nuclei of the joined atoms and the angles between them represent interatomic bond angles. The models may be easily compared in this instance, by placing one on top of the other. For more complex molecules the models cannot be superimposed and planar drawings are best.

Figure 2-1d illustrates how the molecules of H_2S and H_2O may be compared in an orthogonal coordinate system. The heteroatoms have been placed at the origin with one H atom of each molecule along the $+X$ axis. The other H atom is in the $X-Y$ plane. It is now easy to compare the interatomic distances and angles.

The programs of this chapter facilitate the use of x-ray crystallographic coordinate data for building molecular models and for comparing the structures of bioorganic compounds. Orthogonal coordinates are provided for a molecule in a defined orientation. If models are not to be built, or molecules compared, the programs may be simply used to convert x-ray crystallographic data to orthogonal coordinates, distances, and angles without changing the orientation of the molecule. When used to reorient the molecule, the new coordinates may be used to provide a position map against which a molecular model may be compared and for the calculation of interatomic distances and angles. The author has described a technique of building color-coded skeletal models of enzyme active sites using data derived from such transformed coordinates [2,3].

B. Coordinate Transformations

Changing from coordinate data derived by x-ray crystallography to orthogonal coordinates for a molecule in a defined orientation may require a number of transformations of the original coordinate axes. The following must be considered: scale, transformation to an orthogonal coordinate system, rotation, and translation to the defined orientation.

Scale. The original coordinates are each multiplied by a scale factor to provide data that are convenient for model building or for plotting on coordinate paper.

Transformation of axes. X-Ray crystallographic axes are often not mutually perpendicular. The dimensions along each crystallographic axis are proportional to the lengths of the edges of a unit crystal and are usually different for each axis. Thus, the transformation from x-ray crystallographic axes to orthogonal axes must take into account the angles between the axes and the unit dimensions along each axis (Fig. 2-2a).

Translation. The orthogonal axes are moved so that the atom B lies at the origin (Fig. 2-1c).

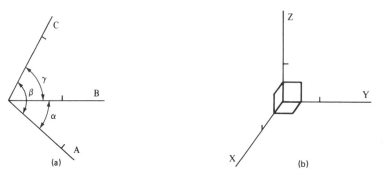

Fig. 2-2 Transformation of axes. (a) X-Ray crystallographic axes are often not mutually perpendicular; (b) an orthogonal coordinate system has its axes mutually perpendicular.

Rotation. Rotation angles are determined as the angles through which the orthogonal axes must move to place the molecule in its standard orientation (Fig. 2-1c).

C. Transformation Equations

All of these changes of scale, transformation of axes, rotation, and translation can be accomplished with three transformation equations, Eqs. (2-1), (2-2), and (2-3).

$$x = \alpha_1 X_f + \beta_1 Y_f + \gamma_1 Z_f + \delta_1 \qquad (2\text{-}1)$$
$$y = \alpha_2 X_f + \beta_2 Y_f + \gamma_2 Z_f + \delta_2 \qquad (2\text{-}2)$$
$$z = \alpha_3 X_f + \beta_3 Y_f + \gamma_3 Z_f + \delta_3 \qquad (2\text{-}3)$$

In these equations X_f, Y_f, and Z_f are the fractional distances along the x-ray crystallographic axes and x, y, and z are the transformed orthogonal coordinates. Program 21 provides the 12 coefficients for the transformation equations.

To obtain the coefficients, the unit distances along the x-ray crystallographic axes are first multiplied by the scale factor. Then orthogonal coordinates are calculated for the key atoms A, B, and C. The coordinates of B are subtracted from those of A and C to place the origin of the new coordinate system at the atom B. Finally, rotation angles are calculated that will place the molecule in its standard orientation. The results of all of these operations are combined to provide the coefficients for the transformation equations.

D. Distances and Angles

Once the 12 coefficients have been calculated, it is relatively simple to transform a set of x-ray crystallographic coordinate data to orthogonal coordinates of a molecule in a defined orientation. The new coordinates may be used to find distances between any two atoms, the angle between any three atoms and the dihedral angle between any four consecutive atoms. Dihedral angles are especially important in model building. Since the bond angles and bond distances are usually incorporated in the model building components, it is only necessary to know the dihedral angles to build a model of a particular conformation of a nonrigid compound. In a structure of four consecutive atoms, ABC and D (Fig. 2-3a), the dihedral angle is the angle formed between AB and CD when viewed along BC. The angle is the same whether the view is from B to C or from C to B. It is negative if movement from D to A is in a counterclockwise direction and positive if this motion is clockwise. The angle lies between 0 and $\pm 180°$ (see Section V-H).

E. Programs

Programs 21 and 22 are designed to facilitate the construction of molecular models from x-ray coordinate data and to provide a convenient means of comparing the structures of bioorganic compounds. Program 21 provides the coefficients for three equations used for coordinate transformations and then provides the new coordinates for any atom as well as the distances between pairs of atoms. Program 22 provides distances between atoms as well as the angles and dihedral angles between atoms that are

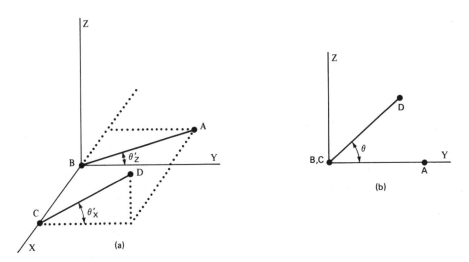

Fig. 2-3 (a) Perspective view of the positive dihedral angle ABCD. θ'_X is the dihedral angle. $\theta'_Z + 90°$ is the angle ABC. (b) Positive dihedral angle ABCD viewed along BC.

useful for molecular model construction. When used in conjunction with program 21, it also provides transformed coordinates of atoms.

II. CALCULATIONS

A. Raw Data

The following data are entered sequentially and used to calculate the coefficients of the transformation equations: the scale; the original coordinates of the atom B at the origin of the coordinate system, the x-ray crystallographic constants a, b, c (dimensions of the unit cell), and α, β, and γ (angles between the crystallographic axes), and, finally, the original coordinates of the atoms A and C. If the orientation of the molecules is not to be defined, then only the scale and x-ray crystallographic parameters are entered.

B. Rotation Angles

The rotation angles (θ_X, θ_Y, and θ_Z) are the angles through which the coordinate system must be rotated about the original axes X, Y, and Z, respectively, to place the molecule in its standard orientation. (These angles differ from the Eulerian angles of classical mechanics [10] by the manner in which the rotations are carried out.) The final position of the molecule with respect to the orthogonal coordinate axes is defined by the *standard angles* θ_{XS}, θ_{YS}, and θ_{ZS} (Fig. 2-1c). The bond BA is placed along the X axis in the positive or negative direction. If the X coordinate of A is positive, θ_{ZS} is 0° and θ_{YS} is 90°. If the X coordinate of A is negative, θ_{ZS} is 180° and θ_{YS} is −90°.

θ_{XS} may have any convenient value. It will be 0° if the atom C lies in the X–Y plane and the Y coordinate of C is positive (as depicted in Fig. 2-1c). To calculate the rotation angles, one first finds the angles through which the bonds BA and BC must move within the orthogonal system to place the molecule in its final position. For convenience of discussion, these angles will be termed *prime angles* (θ'_X, θ'_Y, and θ'_Z) (Fig. 2-4).

To find the rotation angles the prime angles are subtracted in turn from the respective standard angles.

$$\theta_X = \theta_{XS} - \theta'_X, \qquad \theta_Y = \theta_{YS} - \theta'_Y, \qquad \theta_Z = \theta_{ZS} - \theta'_Z$$

Before the prime angles can be calculated, the coordinates of A and C must be expressed in an orthogonal coordinate system in which each axis

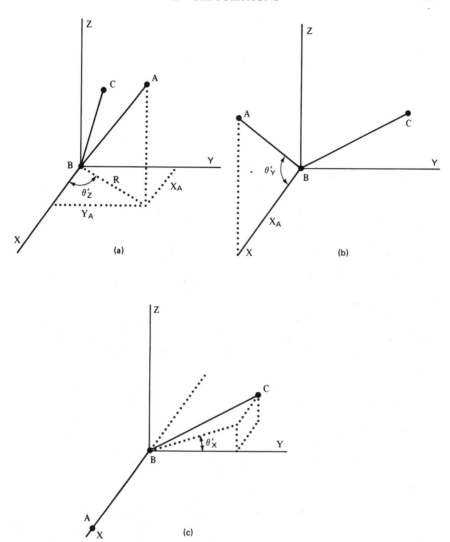

Fig. 2-4 Angle calculations: (a) R moves through the angle θ'_Z to reach the $+X$ axis; (b) BA moves through the angle θ'_Y to reach the $+X$ axis; (c) BC moves through the angle θ'_X in the YZ plane to reach the X–Y plane.

has the same dimension and the atom B is at the origin. The first operation is the change of scale in which the unit cell dimensions a, b, and c are multiplied by the scale factor. Next, the x-ray crystallographic coordinates of A, B, and C are converted to orthogonal coordinates using

$$\begin{bmatrix} X_0 \\ Y_0 \\ Z_0 \end{bmatrix} = \begin{bmatrix} A_1 & B_1 & C_1 \\ 0 & B_2 & C_2 \\ 0 & 0 & C_3 \end{bmatrix} \begin{bmatrix} X_f \\ Y_f \\ Z_f \end{bmatrix} \tag{2-4}$$

where

$$A_1 = a, \qquad C_1 = c \cos \beta$$
$$B_1 = b \cos \gamma, \qquad C_2 = c(\cos \alpha - \cos \gamma \cos \beta)/\sin \gamma$$
$$B_2 = b \sin \gamma, \qquad C_3 = V/ab \sin \gamma$$

where V is the volume of the unit cell:

$$V = abc(1 - \cos^2 \alpha - \cos^2 \beta - \cos^2 \gamma + 2 \cos \alpha \cos \beta \cos \gamma)^{1/2}$$

α, β, and γ are the angles between the crystallographic axes and a, b, and c are the dimensions of the unit cell [11].

Equation (2-4) is applied first to the coordinates X_f, Y_f, and Z_f of the atom A and then to those of C. Finally, the coordinates X_f, Y_f, and Z_f of B are subtracted from the transformed coordinates of A and C. The new coordinates A, B, and C are in an orthogonal system and are referred to with the subscript o.

To find the prime angles through which the molecule must be rotated about each axis in turn to reach the standard orientation defined by θ_{XS}, θ_{YS}, and θ_{ZS}, the rotations are carried out in reverse order: First about the Z axis to give θ'_Z, then about the Y axis so that the atom A lies along the X axis to give θ'_Y, and finally about the X axis to give θ'_X (Fig. 2-4). The angles θ'_X, θ'_Y, and θ'_Z are obtained by using the INV P/R transformation.

To find θ'_Z, X_A is placed in the t register and Y_A in the display. The transformation INV P/R then provides the angle θ'_Z in the display (Fig. 2-4a). (See also Section V.F and Fig. 2-10.) Then θ'_Z is subtracted from θ_{ZS} to provide the angle θ_Z about which the coordinate system is moved to place the molecule ABC in its standard orientation.

After rotation about the Z axis (Fig. 2-4b), A lies in the X–Z plane. The distance R of Fig. 2-4a becomes X_A of Fig. 2-4b. This X_A is still in the t register. Z_A is now recalled from program memory and placed in the display. The operation INV P/R then gives the angle θ'_Y. This angle is subtracted from the standard angle θ_{YS} to provide the angle θ_Y about which the coordinate system is rotated around the Y axis to place ABC in the standard orientation.

θ'_X is the angle through which the molecular ABC must be moved by rotation about the X axis to reach its final orientation after the initial rotations θ_Z and θ_Y have been carried out. To obtain θ'_X, the bond BC is rotated through the angles θ_Z and θ_Y in turn. θ'_X is found in much the same manner as θ'_Z and θ'_Y by using the INV P/R transformation, Section V.F. θ_X is then obtained by subtracting θ'_X from θ_{XS}.

C. Coefficients

The coefficients of Eqs. (2-1), (2-2), and (2-3) will contain the transformations of scale, translation, rotation, and x-ray to orthogonal coordinate change. The latter change is expressed in Eq. (2-4). The scale change was performed in the very beginning of the program. There remains the changes of translation and rotation.

The rotational transformation involves rotation of the coordinate system about each axis Z, Y, and X in turn by the rotation angles θ_Z, θ_Y, and θ_X, respectively. This change can be expressed in matrix form as

$$\begin{bmatrix} X \\ Y \\ Z \end{bmatrix} = \begin{bmatrix} a_1 & b_1 & c_1 \\ a_2 & b_2 & c_2 \\ a_3 & b_3 & c_3 \end{bmatrix} \begin{bmatrix} X_R \\ Y_R \\ Z_R \end{bmatrix} \tag{2-5}$$

$a_1 = \cos \theta_Y \cos \theta_Z$

$a_2 = \cos \theta_X \sin \theta_Z + \sin \theta_X \sin \theta_Y \cos \theta_Z$

$a_3 = \sin \theta_X \sin \theta_Z - \cos \theta_X \sin \theta_Y \cos \theta_Z$

$b_1 = -\cos \theta_X \sin \theta_Z$

$b_2 = \cos \theta_X \cos \theta_Z - \sin \theta_X \sin \theta_Y \sin \theta_Z$

$b_3 = \sin \theta_X \cos \theta_Z + \cos \theta_X \sin \theta_Y \sin \theta_Z$

$c_1 = \sin \theta_Y$

$c_2 = -\sin \theta_X \cos \theta_Y$

$c_3 = \cos \theta_X \cos \theta_Y$

Equation (2-5) is very similar to the corresponding expression derived for Eulerian angles [10]. The difference reflects the order in which the rotations are carried out.

To obtain the α, β, γ, and δ coefficients, the expression

$$\begin{bmatrix} a_1 & b_1 & c_1 \\ a_2 & b_2 & c_2 \\ a_3 & b_3 & c_3 \end{bmatrix}$$

is multiplied in turn by the following expressions, respectively,

$$\begin{bmatrix} A \\ 0 \\ 0 \end{bmatrix} ; \begin{bmatrix} B_1 \\ B_2 \\ 0 \end{bmatrix} ; \begin{bmatrix} C_1 \\ C_2 \\ C_3 \end{bmatrix} ; \begin{bmatrix} -X_B \\ -Y_B \\ -Z_B \end{bmatrix}$$

using the matrix multiplication capacity of the Master Library Module. An excellent discussion of this use of matrices is provided by Ledley [12].

The first three expressions are components of the x ray to orthogonal coordinate transformation and the last one is the translation operation. The results of these matrix multiplications may be expressed in algebraic form as follows:

$$\alpha_1 = a_1A_1; \qquad \beta_1 = a_1B_1 + b_1B_2$$
$$\alpha_2 = a_2A_1; \qquad B_2 = a_2B_1 + b_2B_2$$
$$\alpha_3 = a_3A_1; \qquad \beta_3 = a_3B_1 + b_3B_2$$
$$\gamma_1 = a_1C_1 + b_1C_2 + c_1C_3$$
$$\gamma_2 = a_2C_1 + b_2C_2 + c_2C_3$$
$$\gamma_3 = a_3C_1 + b_3C_2 + c_3C_3$$
$$\delta_1 = a_1D_1 + b_1D_2 + c_1D_3$$
$$\delta_2 = a_2D_1 + b_2D_2 + c_2D_3$$
$$\delta_3 = a_3D_1 + b_3D_2 + c_3D_3$$

D. Coordinate Transformations

Equations (2-1), (2-2), and (2-3) employ the coefficients described above to carry out the coordinate transformations. The equations are so simple that it is only necessary to enter the original x-ray (or other) coordinate data and to use the stored coefficients with simple arithmetic to provide the final coordinates x, y, and z.

E. Distances

The distance between two atoms A and B in three-dimensional orthogonal space is given by Eq. (2-6).

$$D = [(x_A - x_B)^2 + (y_A - y_B)^2 + (z_A - z_B)^2]^{1/2} \qquad (2\text{-}6)$$

F. Angles

An *angle* ABC is obtained by placing the three atoms ABC on the $X-Y$ plane of an orthogonal coordinate system with B at the origin and using the INV P/R transformation to provide the angle θ'_z (Fig. 2-3a). The angle ABC is θ'_z plus 90°.

G. Dihedral Angles

A *dihedral angle* involving four consecutive atoms ABCD is obtained by placing ABC on the $X-Y$ plane with B at the origin as in Section II.F and then obtaining the angle θ'_x, which the bond CD makes with the $X-Y$ plane (Fig. 2-3b). When a series of dihedral angles are calculated along a chain of atoms as in Fig. 2-5, successive angles may involve branching or a continuation along the chain of atoms. The difference becomes important in program design. When using the program, only the coordinates of the new atom need to be added if those of the other three atoms have already been entered (see Section V.G).

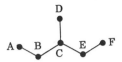

Fig. 2-5 Dihedral angles along a chain of atoms: (a) ABCD is measured first; (b) ABCE then becomes a branch; (c) BCEF is now a continuation.

III. EXAMPLES

Five examples have been selected to illustrate the scope and variety of the problems that may be solved with coordinate transformations. The first example illustrates how program 21 can be used to compare two molecules that interact with the same biological receptor. The second and third examples illustrate how x-ray crystallographic data may be used directly to provide distances, angles, and dihedral angles. Small molecules have been selected, which crystallize in the triclinic system. The fourth example provides x-ray crystallographic data for a small peptide chain, in this case, a tripeptide. Program 22 is used to provide rotation angles so that a molecular model may be built without even constructing a graph to locate the atoms. The final example describes the use of programs 21 and 22 to provide data for the construction of a model of the active site of an enzyme.

A. The Structural Relationship of Phorbol and Cortisol

Two compounds (Fig. 2-6, Table 2-1) may be compared in three-dimensional space by using a least-squares technique in which the molecules are orientated so that the distances between comparable atoms are minimized [13]. The calculation requires a computer. More elaborate, computer-based methods allow the visualization of the two molecules in three dimensions as comparable features are brought as close together as possible [1].

Program 21 provides a much simpler method of making the comparison. Three comparable atoms of the two compounds are selected. One pair of comparable atoms is placed at the origin of an orthogonal coordinate system. A second pair is made to lie along the X axis. The third pair is placed in a convenient orientation by rotating the molecules about the X axis. One such orientation, for instance, is to place the third pair of comparable atoms in the $X-Y$ plane. The exact location of the second and third pairs of atoms can now be compared to provide a measure of how closely the two molecules can be made to coincide. The locations of other comparable pairs of atoms or features can also be examined.

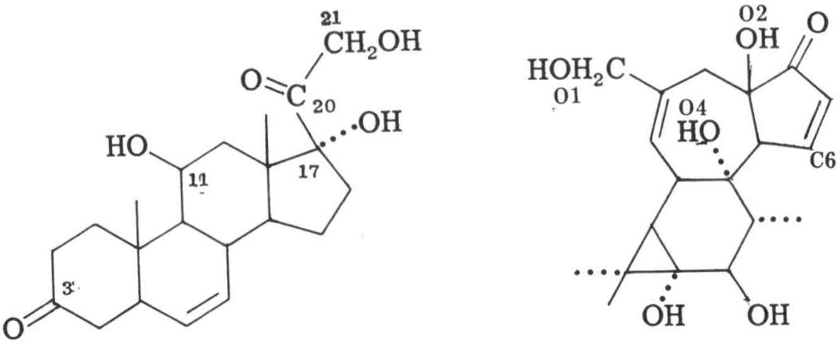

Fig. 2-6 Structures of (a) cortisol and (b) phorbol.

This example compares two molecules of very different structures. One compound, cortisol, is a hormone in mammalian cells associated with cell growth. The second compound, phorbol, is a plant product that promotes tumor growth in mouse skin. Although the two molecules are structurally very different, Wilson and Huffman [13] have proposed that phorbol may bind to a cortisol receptor in such a way as to alkylate the receptor. The authors suggest that some group on the receptor, such as an amino or sulfhydryl residue, normally forms a hydrogen bond with the C-20 carbonyl of cortisol. However, when phorbol interacts with the re-

TABLE 2-1

X-Ray Crystallographic Parameters of Cortisol[a] and Phorbol[b]

Atom[c]	X/a	Y/b	X-c	
Cortisol				
O-21	0.7193	0.0802	0.1309	$a = 14.372$ Å
O-11	0.4961	0.3764	0.1516	$b = 18.400$ Å
O-17	0.8073	0.2726	−0.0883	$c = 7.706$ Å
O-20	0.6535	0.128	0.1629	$\alpha = \beta = \gamma = 90°$
Phorbol				
O-4	−0.0387	0.0197	0.5501	$a = 10.54$ Å
O-1	−0.2322	0.3933	0.1715	$b = 10.35$ Å
O-2	−0.3895	0.2364	0.4782	$c = 13.80$ Å
C-6	−0.1131	0.2681	0.6418	$\alpha = 90°$
				$\beta = 107.6°$
				$\gamma = 90°$

[a] Reference 14.
[b] Reference 16.
[c] See Fig. 2-6 for atom designations.

ceptor, there is presented a cyclopentenone system at the same location as the C-20 carbonyl of cortisol would occupy. In fact, the β carbon of the cyclopentenone system is located so that the amino or sulfhydryl residue will interact with it by a conjugate addition reaction. The ORTEP stereopair drawings provided by Wilson and Huffman show how the two molecules can be made to have corresponding crucial function groups overlap. The results of the least-squares fit shows that the distances between comparable pairs of atoms are 0.263–0.577 Å [6a].

The critical functional groups are O-21, O-11, O-17, and O-20 for cortisol and O-4, O-1, O-2, and C-6, respectively, for phorbol (Fig. 2-6). Table 2-1 lists the pertinent x-ray crystallographic data for the two compounds [14–16]. Crystals of cortisol are orthorhombic, while phorbol (as its 12-myristate, 13-acetate diester) forms monoclinic crystals.*

Table 2-2 is the calculator printout using programs 211, 212, 213, and 214. Columns 1 and 2 provide printouts of the raw data as entered and columns 3 and 4 are the results. The data entry are as follows for cortisol:

28–scale: The results are desired in Å so the scale is not changed.

29, 30, 31: These are the fractional coordinates of the atom at the origin, in this case O-21.

32, 33, 34: These are the unit cell dimensions, a, b, and c in angstroms.

35, 36, 37: These are the angles between the crystallographic axes, in this case all are 90° since the crystal is orthorhombic.

38, 39, 40: These angles provide a standard orientation. Since A is along the $+X$ axis, θ_{ZS} and θ_{YS} are 0° and 90°, respectively. Since C is in the $X–Y$ plane in the $+Y$ direction, θ_{XS} is 0°.

41, 42, 43: These are the fractional coordinates of the atom A, O-11, that is placed along the X axis.

44, 45, 46: Fractional coordinates of the atom C in the $X–Y$ plane, in this case O-17.

The key A is pressed and, after about 1.5 min, the angles θ_Z, θ_Y, and, finally after a pause, θ_X are printed.

Programs 213 and 214 are now entered. The key A′ is pressed to complete the calculation. Nothing is printed, but a number appears in the display when calculation is complete. B′ is pressed to print the coefficients at the top of column 3.

In this example, operation proceeded directly to obtain the appropriate transformed coordinates. O-21 was placed at the origin so its coordinates are confirmed to be zero. O-11 was placed along the X axis. The distance of O-11 from O-21 is seen to be 6.33 Å. O-17 is in the $X–Y$ plane. Its X coordinate is 2.37 Å and its Y coordinate is 3.37 Å.

* Reference 15, quoted by Wilson and Huffman [13] actually provides the coordinates of cortisone; for our example we selected the coordinates of cortisol, ref. 14.

TABLE 2-2

Comparison of Cortisol and Phorbol

Data input[a]		Results	
Cortisol	Phorbol	Coefficients	Phorbol
28.	28.	-7.287784651	4. □
1.	1.	10.49626349	-0.0387
29.	29.	-6.578072038	0.0197
0.7193	-0.0387	15.85212995	0.5501
30.	30.	8.18849664	0.00
0.0802	0.0197	-4.496498502	0.00
31.	31.	.1943100567	0.00
0.1309	0.5501	-3.993560132	1. □
32.	32.	-6.587575963	-0.2322
14.372	10.54	3.945327491	0.3933
33.	33.	-7.683922734	0.1715
18.4	10.35	5.95454009	6.32
34.	34.		0.00
7.706	13.8		0.00
35.	35.	Cortisol	2. □
90.	90.		-0.3895
36.	36.	21. □	0.2364
90.	107.6	0.7193	0.4782
37.	37.	0.0802	2.36
90.	90.	0.1309	3.45
38.	38.	0.00	0.00
0.	0.	0.00	6. C
39.	39.	0.00	-0.1131
90.	90.	11. □	0.2861
40.	40.	0.4961	0.6418
0.	0.	0.3764	0.82
41.	41.	0.1516	2.05
0.4961	-0.2322	6.33	2.35
42.	42.	0.00	
0.3764	0.3933	0.00	
43.	43.	17. □	Distance
0.1516	0.1715	0.8073	C - Cortisol to
44.	44.	0.2726	C - 6 Phorbol
0.8073	-0.3895	-0.0883	
45.	45.	2.37	1.18
0.2726	0.2364	3.37	0.87
46.	46.	0.00	2.15
-0.0883	0.4782	20. □	0.71
47.	47.	0.6535	2.01
		0.128	2.21
		-0.1629	1.23 - distance
-120.48039	-96.77985029	1.18	
1.444890511	-51.97788057	0.87	
-211.2253349	-25.57553894	2.15	

[a] Data from Table 2-1.

The data for phorbol are handled just as for cortisol. It is evident that O-1 is nearly coincident with O-11 of cortisol and O-2 is very close to O-17 of cortisol.

It is necessary to press D′ and reenter the transformed coordinates of O-20 of cortisol and C-6 of phorbol to calculate the distance, 1.23 Å,

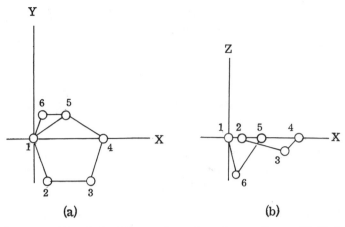

Fig. 2-7 Structure of bicyclo-3,1,0-hexane drawn from the coordinates of Table 2-4: (a) in the $X-Y$ plane; (b) in the $X-Z$ plane.

between these atoms. A group, such as sulfhydryl, which is within hydrogen-bonding distance of the O-20 of cortisol, is certainly very close as well to the enone system of phorbol. Thus, the proposal of Wilson and Huffman [13] seems reasonable on structural grounds, although in other respects the molecules are very dissimilar.

B. Bicyclo[3,1,0]hexane

Only six atoms are included in this example (Fig. 2-7, Table 2-3) although the original reference [17] provides the coordinates of a substituent. This example is thus perhaps the simplest test of the proper working of the two programs, 21 and 22. Three of the six atoms are used

TABLE 2-3

X-Ray Crystallographic Parameters of Bicyclo[3,1,0]hexane[a]

Atom[b]	X/a	Y/b	Z/c	
1	0.8858	0.1890	−0.3366	$a = 9.89$ Å
2	0.9896	0.3154	−0.3893	$b = 9.53$ Å
3	1.1008	0.3378	−0.2080	$c = 5.602$ Å
4	1.0992	0.1864	−0.1641	$\alpha = 104.71°$
5	0.9523	0.1108	−0.1980	$\beta = 83.59°$
6	0.8397	0.1980	−0.0671	$\gamma = 100.20°$

[a] Reference 7.
[b] See Fig. 2-7 for atom designations.

to orient the molecule. The other three are then added so that distances, angles, and dihedral angles may be checked. This example also illustrates the calculation of distances, angles, and dihedral angles without determining the orientation of the molecule.

Atoms 4, 1, and 5 are used for orientation as atoms A, B, and C, respectively. The data are entered into programs 211 and 212 as in Example A, except that the scale of 1.25 cm/Å is used. After A' is pressed in programs 213 and 214 and the coefficients are calculated, B' is pressed to print these out (top of column 2, Table 2-4).

Operation then went directly to programs 221, 222, and 223. Column 2 of Table 2-4 illustrates calculation of distances and angles. With program 22 rectangular coordinates are printed to four places of decimals. The number labels punctuate the coordinate sets. The first entry of column 2 shows that atom 1 is at the origin. Atom 2 is -0.0164 cm from the plane of atoms 1, 4, and 5 since these are in the $X-Y$ plane ($Z = 0$) and the Z coordinate of atom 2 is -0.0164. The label 102 indicates that the distance between atoms 1 and 2 has been calculated; this distance is 1.88 cm.

Column 2 then contains the raw data for calculation of the angle between atoms 2, 1, and 5. These had to be reentered in this instance because the coordinates of atoms 1 and 2 had been entered in the wrong order to proceed with calculating the angle 2, 1, 5. The angle 2, 1, 5 is 107.9°, identical with the value reported in the original reference.

Columns 3 and 4 of Table 2-4 illustrate the calculation of dihedral angles. The first one calculated is 2-1-6-5. The next one, 1-6-5-4, is a continuation along the chain of atoms so that only the coordinates of the new atom 4 need to be entered. The calculation of successive dihedral angles continues around the ring until 3-2-1-5 is calculated. The next angle 3-2-1-6, involves a branch, rather than a continuation along the chain. Accordingly, label B' was pressed before the data for atom 6 were entered. This procedure of calculating all the dihedral angles provided the rectangular coordinates as well, and it was not necessary to use a separate operation for these.

Of course, if the original coordinates are already in an orthogonal coordinate system, it is not necessary that they be transformed to calculate distances, angles, or dihedral angles. The last column of Table 2-4 shows what the printout looks like when the input data are orthogonal. The angle 2-1-5 is again found to be 107.9°. It will be shown in the next example that the input of raw data correct to only two decimal places does not always provide the same angle as data correct to four decimal places. The difference arises from rounding off during the calculations.

Table 2-5 is the calculator printout when the data of Table 2-3 are used to calculate orthogonal coordinates without determining the orientation of

TABLE 2-4

Structural Parameters of Bicyclo[3,1,0]hexane[a]
Calculator Printout[b]

Data input	Coefficients	Dihedral angles		Angles
28.	11.33728216	2.	2.	2.
1.25	-3.02375018	0.9896	0.9896	0.5928
29.	3.892861968	0.3154	0.3154	-1.7862
0.8858	-3.164640394	-0.3893	-0.3893	-0.0164
30.	-10.15242059	0.5928	0.5928	
0.189	-5.36852527	-1.7862	-1.7862	
31.	3.490730635	-0.0164	-0.0164	1.
-0.3366	3.587663739	1.	5040302.	0.
32.	-4.896782085	0.8858	-29.3	0.
9.89	-8.269467572	0.1890		0.
33.	5.804853016	-0.3366		
9.53	-4.081902705	0.0000	1.	5.
34.		0.0000	0.8858	1.4852
5.602	Distance	0.0000	0.1890	1.0901
35.	1.	6.	-0.3366	0.
104.71	0.8858	0.8397	0.0000	
36.	0.1890	0.1980	0.0000	
83.59	-0.3366	-0.0671	0.0000	20105.
37.	0.0000	0.3896	4030201.	107.9
100.2	0.0000	1.0149	29.2	
38.	0.0000	-1.5475		5.
0.	2.	5.	5.	1.4852
39.	0.9896	0.9523	0.9523	1.0901
90.	0.3154	0.1108	0.1108	0.
40.	-0.3893	-0.1980	-0.1980	
0.	0.5928	1.4852	1.4852	1.
41.	-1.7862	1.0901	1.0901	0.
1.0992	-0.0164	0.0000	0.0000	0.
42.	102.	2010605.	3020105.	0.
0.1864	1.88	-97.0	-17.8	
43.				6.
-0.1641	Angle	4.	6.	0.3896
44.	2.	1.0992	0.8397	1.0149
0.9523	0.9896	0.1864	0.1980	-1.5475
45.	0.3154	-0.1641	-0.0671	
0.1108	-0.3893	3.0298	0.3896	50106.
46.	0.5928	0.0000	1.0149	61.1
-0.198	-1.7862	0.0000	-1.5475	
47.	-0.0164	1060504.	3020106.	
	1.	97.2	48.1	
	0.8858			
	0.1890	3.		
6.526465088	-0.3366	1.1008		
22.62354424	0.0000	0.3378		
-158.7246587	0.0000	-0.2080		
	0.0000	2.4155		
	5.	-1.6994		
	0.9523	-0.5916		
	0.1108	6050403.		
	-0.1980	-47.1		
	1.4852			
	1.0901			
	0.0000			
	20105.			
	107.9			

[a] Data from Table 2-3.
[b] See example Section III.B for explanation.

TABLE 2-5

Structural Parameters of Bicyclo[3,1,0]hexane[a]
Calculator Printout when Orientation is not Changed[b]

Data input	Distance	Dihedral angles
28.	1.	2.
1.	0.8858	0.9896
29.	0.1890	0.3154
32.	-0.3366	-0.3893
9.89	6.5127	7.7321
33.	-0.2191	0.5961
9.53	5.7907	6.1304
34.	2.	1.
5.602	0.9896	0.8858
35.	0.3154	0.1890
104.71	-0.3893	-0.3366
36.	7.7321	6.5127
83.59	0.5961	-0.2191
37.	6.1304	5.7907
100.2	102.	6.
38.	1.51 (Angstroms)	0.8397
		0.1980
-45.		-0.0671
-45.		5.1940
-45.	1.88 (centimeters)	0.3190
		6.3010
	Angle	5.
Coefficients	2.	0.9523
	0.9896	0.1108
4.945	0.3154	-0.1980
-1.448356967	-0.3893	5.9620
8.441643033	7.7321	-0.7566
3.845883284	0.5961	7.0477
8.25295101	6.1304	2010605.
-2.81405071	1.	-97.0
-4.175662012	0.8858	
1.473324365	0.1890	4.
3.43155447	-0.3366	1.0992
0.	6.5127	0.1864
0.	-0.2191	-0.1641
0.	5.7907	6.8376
	5.	-0.2955
	0.9523	8.1914
	0.1108	1060504.
	-0.1980	97.2
	5.9620	
	-0.7566	
	7.0477	
	20105.	
	107.9	

[a] Table 2-3.
[b] See Sections III.B for explanation.

the molecule. In this case, data input is simplified. The scale chosen, 1, is that of the original authors so that the coordinates and dimensions will be in angstroms. Original coordinates of the orientation atoms A, B, and C are not required, so label B' is pressed and only the x-ray crystallographic parameters are entered. The rotation angles, in this instance, each appear as −45°. The equation coefficients, printed at the bottom of column 1 have zero values for the δs since the origin of the coordinate system is not changed.

The data input and the printout with program 22 appear as in Table 2-4 except that now none of the atoms is at the origin of the coordinate system. The distance between atoms in angstroms must be multiplied by the scale factor of 1.25 to provide the same result as in Table 2-4. The angle calculations are the same as in Table 2-4.

C. Maleic Hydrazide (1,2-Dihydropyridazine-3,6-dione)

In spite of its name, x-ray crystallographic data [18] prove that only one of the oxygens of this compound (Fig. 2-8, Table 2-6) is a carbonyl in the crystal. This example adds very little to that of the example in Section III.B, but it is included to illustrate all the labels of program 21. Our program labels easily accommodate 2-H and 3-H but have no designation for H(O-2), i.e., for the hydrogen atom attached to oxygen atom number 2. An arbitrary designation of 1 OH and 2 OH was used here but the reader may find it more convenient to renumber all the atoms in one consecutive sequence to avoid this problem.

The output for this example (Tables 2-7 and 2-8) shows, as the original authors pointed out, that maleic hydrazide is a nearly planar molecule. This is shown by the fact that, in our orientation of the molecule, all the Z coordinates are nearly zero. One atom, the hydrogen attached to N-1, is out of the plane by 0.20 cm (0.16 Å). The last data set in column 4 illustrates a distance calculation with program 21. The same calculation is repeated in Table 2-8, this time with program 22, to show the advantage of

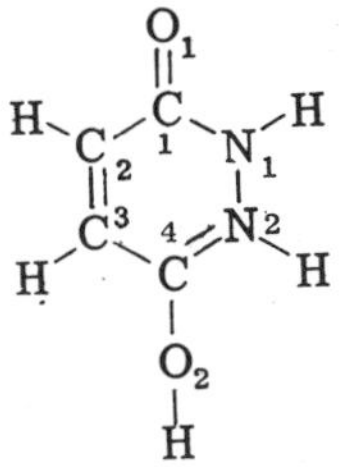

Fig. 2-8 Structure of maleic hydrazide (1,2-dihydropyridazine-3,6-dione).

TABLE 2-6

X-Ray Crystallographic Parameters for Maleic Hydrazide[a]

Atom[b]	X/a	Y/b	Z/c	
O-1	0.9667	1.2598	0.3551	$a = 5.83$ Å
O-2	0.0743	0.7087	0.1831	$b = 5.78$ Å
N-1	0.6785	0.9017	0.3691	$c = 7.31$ Å
N-2	0.4547	0.7465	0.3333	$\alpha = 79.0°$
C-1	0.7528	1.1410	0.3099	$\beta = 99.5°$
C-2	0.5718	1.2427	0.1931	$\gamma = 107.2°$
C-3	0.3485	1.1019	0.1520	
C-4	0.2979	0.8482	0.2260	
H(C-2)	0.627	1.419	0.160	
H(C-3)	0.199	1.147	0.066	
H(N-1)	0.789	0.855	0.459	
H(O-2)	0.048	0.549	0.226	

[a] Reference 18.

[b] See Fig. 2-8 for atom designations.

labeled data. Angles and dihedral angles were calculated also, as shown in Table 2-8, using the data of program 21 (Table 2-7). The results are not as accurate as they would be if the original coordinates, correct to four decimal places, are used. The value 120.9° for the angle 1-2-3 obtained with the original coordinates is identical with the result reported by the original authors [18].

D. Sarcosylglycylglycine—a Tripeptide

This example [19] was selected to illustrate the calculation of dihedral angles for a peptide chain for use in model building (Fig. 2-9, Table 2-9). With the dihedral angles and a few selected interatomic distances, models of peptide chains such as this one may be easily constructed [2,3].

Column 1 of Table 2-10 is the printout from program 21: first the raw data input and then the rotation angles. In this case, R_{06} was recalled to provide the volume of the unit crystal. Since the scale is 1.25 cm/Å, this value is divided by 1.25^3 to give the volume of the unit crystal in angstroms. The value is the same as provided by the original authors (Table 2-9). The coefficients are printed at the top of column 2. C-12 was selected as the atom to be placed at the origin of the coordinate system. Negative X values are avoided by placing C-2 along the $+X$ axis. One of the carbonyl oxygens, O-5, was placed in the $X-Y$ plane. The printout at the bottom of column 2 confirms that the programs are operating, since

TABLE 2-7

Structural Parameters of Maleic Hydrazide

Data input[a]	Coefficients	Atomic coordinates	Atomic coordinates	
28.	-5.553927545		2. C	20. H
1.25	3.025791983		0.5718	0.048
29.	-3.620238659		1.2427	0.549
0.9667	-2.547015377		0.1931	0.226
30.	-5.925879605		2.52	7.14
1.2598	3.255347689		-1.61	1.02
31.	-1.727597533		0.02	-0.07
0.3551	3.217988904		3. C	
32.	8.375790138		0.3485	
5.83	9.191181613		1.1019	
33.	3.397682157		0.152	
5.78	-3.575645384		4.19	1.57
34.			-1.59	-0.09
7.31			0.02	0.01
35.	Atomic		4. C	2.52 C
79.	coordinates		0.2979	-1.61
36.	1. O		0.8482	0.02
99.5	0.9667		0.226	1.79
37.	1.2598		4.99	
107.2	0.3551		0.00	
38.	0.00		0.00	
0.	0.00		2. H	
39.	0.00		0.627	
90.	2. O		1.419	
40.	0.0743		0.16	
0.	0.7087		1.82	
41.	0.1831		-2.60	
0.2979	6.66		0.11	
42.	0.01		3. H	
0.8482	0.00		0.199	
43.	1. N		1.147	
0.226	0.6785		0.066	
44.	0.9017		5.05	
0.4547	0.3691		-2.58	
45.	2.49		-0.01	
0.7465	1.30		10. H	
46.	-0.01		0.789	
0.3333	2. N		0.855	
47.	0.4547		0.459	
	0.7465		1.84	
Rotation angles	0.3333		2.20	
-218.4415287	4.19		0.20	
-13.3376565	1.42			
-33.90599762	0.00			
	1. C			
	0.7528			
	1.141			
	0.3099			
	1.57			
	-0.09			
	0.01			

TABLE 2-8

Structural Parameters of Maleic Hydrazide[a]

Calculator Printout using Program 22

Distance

```
     1.
  1. 57
 -0. 09
  0. 01

     2.
  2. 52
 -1. 61
  0. 02

   102.
  1. 79


     1.
0. 7528
1. 1410
0. 3099
1. 5687
-0. 0887
0. 0090
     2.
0. 5718
1. 2427
0. 1931
2. 5167
-1. 6149
0. 0171
   102.
  1. 80
```

Angle

```
     1.
  1. 57
 -0. 09
  0. 01

     2.
  2. 52
 -1. 61
  0. 02

     3.
  4. 19
 -1. 59
  0. 02


 10203.
 121. 3


     1.
0. 7528
1. 1410
0. 3099
1. 5687
-0. 0887
0. 0090
     2.
0. 5718
1. 2427
0. 1931
2. 5167
-1. 6149
0. 0171
     3.
0. 3485
1. 1019
0. 1520
4. 1865
-1. 5884
0. 0229
 10203.
 120. 9
```

Dihedral angle

```
     3.
  4. 19
 -1. 59
  0. 02

     4.
  4. 99
     0.
     0.

     2.
  6. 66
  0. 01
     0.

     2.
  7. 14
  1. 02
 -0. 07

3040202.
 176. 7

     3.
0. 3485
1. 1019
0. 1520
4. 1865
-1. 5884
0. 0229
     4.
0. 2979
0. 8482
0. 2260
4. 9859
0. 0000
0. 0000
     2.
0. 0743
0. 7087
0. 1831
6. 6571
0. 0120
-0. 0040
     2.
0. 0480
0. 5490
0. 2260
7. 1358
1. 0169
-0. 0693
3040202.
 177. 1
```

[a] Data input from Tables 2-6 and 2-7.

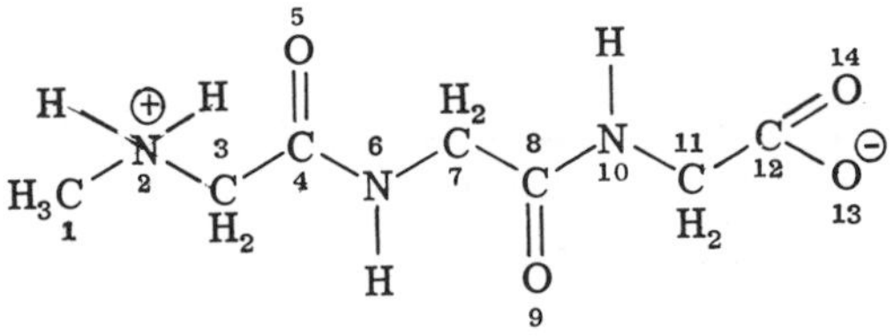

Fig. 2-9 Structure of sarcosylglycylglycine.

C-12 is obviously at the origin, C-2 along the $+X$ axis, and O-5 is in the $X-Y$ plane with a positive value for Y. Column 3 illustrates again the format for distances and angle calculations.

Table 2-11 provides the dihedral angles as one proceeds down the chain from C-1. It is interesting that atoms 1, 2, 3, and 4 lie almost in a plane (rotation angle 176.4°). The dihedral angle about the C-3-C-4 bond is $-24.0°$ when O-5 is included, and 158.5° when C-6 is included. The amide bonds are nearly planar: $-178.2°$ for 3-4-6-7, and $-176.2°$ for 7-8-10-11. The last five atoms N-10, C-11, C-12, O-13, and O-14 are almost in a plane since rotation angles 10-11-12-13 and 10-11-12-14 are 177.5° and $-3.1°$, respectively. The signs of these rotation angles are the opposite of those reported by the original authors. It is possible that the original authors define the sign of the angles differently.

TABLE 2-9

X-Ray Crystallographic Parameters for Sarcosylglycylglycine[a]

Atom[b]	X/a	Y/b	Z/c	
1	−0.2610	−0.1885	0.1475	$a = 20.847$ Å
2	−0.2230	−0.0667	0.1594	$b = 10.252$ Å
3	−0.1683	−0.0787	0.2661	$c = 8.719$ Å
4	−0.1300	0.0473	0.2628	$\alpha = \beta = \gamma = 90°$
5	−0.1318	0.1200	0.1511	$V = 1863.4$ Å^3
6	−0.0932	0.0687	0.3862	
7	−0.0565	0.1879	0.3952	
8	0.0009	0.1974	0.2891	
9	0.0253	0.3048	0.2653	
10	0.0227	0.0868	0.2288	
11	0.0799	0.0841	0.1330	
12	0.1425	0.0733	0.2222	
13	0.1930	0.0659	0.1445	
14	0.1418	0.0733	0.3654	

[a] Reference 19.
[b] See Fig. 2-9 for atom designations.

TABLE 2-10

Structural Parameters of Sarcosylglycylglycine

Data input[a]	Coefficients	Distance
28.	-25.54477035	1.
1.25	-4.503759531	-0.2610
29.	2.497860971	-0.1855
0.1425	-2.366315886	0.1475
30.	12.47962441	10.9771
0.0733	-1.698160356	-1.2923
31.	-.7677511712	0.2348
0.2222	-1.608651425	2.
32.	-10.75200215	-0.2230
20.847	3.984175039	-0.0667
33.	.0844716101	0.1594
10.252	2.157624844	9.7161
34.		0.0000
8.719		0.0000
35.	Atomic	102.
90.	coordinates	1.82
36.	12.	
90.	0.1425	Angle
37.	0.0733	1.
90.	0.2222	-0.2610
38.	0.0000	-0.1885
0.	0.0000	0.1475
39.	0.0000	10.9842
90.	2.	-1.3297
40.	-0.2230	0.2399
0.	-0.0667	2.
41.	0.1594	-0.2230
-0.223	9.7161	-0.0667
42.	0.0000	0.1594
-0.0667	0.0000	9.7161
43.	5.	0.0000
0.1594	-0.1318	0.0000
44.	0.1200	3.
-0.1318	0.1511	-0.1683
45.	6.9510	-0.0787
0.12	1.9326	0.2661
46.	0.0000	8.2653
0.1511		-0.5678
47.		-0.9902
		10203.
		112.7

Rotation angles
-190.6676568
-4.03948753
-188.509142

Unit crystal vol.
3639.559977   cm³
1863.454708   A³

[a] Table 2-9.

TABLE 2-11

Structural Parameters of Sarcosylglycylglycine

X – Ray data input[a] Dihedral angles			Orthogonal input[b] Distance	Orthogonal input[b] Dihedral angles
1.	7.	12.	1.	1.
-0.2610	-0.0565	0.1425	10.9842	10.9842
-0.1885	0.1879	0.0733	-1.3297	-1.3297
0.1475	0.3952	0.2222	0.2399	0.2399
10.9842	4.6794	0.0000		
-1.3297	2.0481	0.0000	2.	2.
0.2399	-2.5518	8101112.	9.7161	9.7161
2.	3040607.	85.5	0.	0.
-0.2230	-178.2		0.	0.
-0.0667				
0.1594	8.	13.	102.	3.
9.7161	0.0009	0.1930	1.85	8.2653
0.0000	0.1974	0.0659		-0.5678
0.0000	0.2891	0.1445	Angle	-0.9902
3.	3.2721	-1.2128	1.	
-0.1683	2.0788	-0.1948	10.9842	4.
-0.0787	-1.2837	0.9741	-1.3297	6.9913
0.2661	4060708.	10111213.	0.2399	0.8375
8.2653	-72.1	177.5		-1.073
-0.5678			2.	
-0.9902	9.	14.	9.7161	1020304.
4.	0.0253	0.1418	0.	176.4
-0.1300	0.3048	0.0733	0.	
0.0473	0.2653	0.3654		5.
0.2628	2.4130	-0.0921	3.	6.951
6.9913	3.3475	-0.2272	8.2653	1.9326
0.8375	-1.1493	-1.5414	-0.5678	0.
-1.0730	6070809.	10111214.	-0.9902	
1020304.	163.7	-3.1		2030405.
176.4			10203.	-24.0
	10.		112.7	
5.	0.0227			6.
-0.1318	0.0868			5.9059
0.1200	0.2288			0.7403
0.1511	3.0233			-2.3443
6.9510	0.6974			
1.9326	-0.3931			2030406.
0.0000	6070810.			158.5
2030405.	-17.4			
-24.0				7.
	11.			4.6794
6.	0.0799			2.0481
-0.0932	0.0841			-2.5518
0.0687	0.1330			
0.3862	1.6420			3040607.
5.9059	0.5602			-178.2
0.7403	0.7844			
-2.3443	7081011.			
2030406.	-176.2			
158.5				

[a] Table 2-9.
[b] Table 2-10.

The last two columns of Table 2-10 illustrate once more the calculation of distances, angles, and rotation angles using program 22 when the raw data input is orthogonal.

The dihedral angles may be used directly to construct a skeletal model of the tripeptide (see ref. 2). The method may be extended to longer peptides quite easily (see also the example in Section III.E).

E. Models of Enzyme Active Sites

Programs 21 and 22 may be used to provide the coordinate and dihedral angle data required to build a model of the active site of an enzyme. The technique is relatively simple and the user can add as much or as little as desired of the molecular architecture around the site. The construction of a model of the active site of α-chymotrypsin was described by the author [3] who used it to study the stereochemistry and mechanism of action of the enzyme. Lawson and Rao [5a] and Lawson [5b] have used this model for the design of enzyme inhibitors.

When the original work on the chymotrypsin-active site model [3] was carried out, the programs of this chapter had not yet been developed. Now with programs 21 and 22 available, the construction of other enzyme-active sites is simplified. The coordinates of many enzymes are available from the Protein Data Bank at Brookhaven National Laboratory [20], and references to descriptions of the enzymes are supplied with the coordinates.

The following steps are involved in the construction of a model of the enzyme-active site.

1.   Model building components are assembled. Although the author prefers color-coded skeletal models [3], these are not commercially available. Nicholson Molecular Models supplied by LABQUIP [21] may be used since these retain fixed dihedral angles.

2.   Three key atoms are selected of the active site or of an inhibitor that is bound at the active site and for which x-ray crystallographic data are available. The three atoms are to be placed in the standard orientation with one of them at the origin of the coordinate system, one along the X axis and the third in the $X-Y$ plane. Program 21 will provide the coefficients for the transformation equations and program 22 will generate labeled, transformed coordinates for each atom in the active site region. Usually these are described by the original authors and depicted in the drawings of their publications.

3.   The coordinates are plotted in the $X-Y$ plane and again in the $X-Z$ plane to provide elevation.

4. The coordinates are used to provide dihedral angles for the amino acid residues of the active site. These are obtained and labeled with program 22.

5. Finally, the coordinates and dihedral angles are used to construct the individual amino acid components and these are assembled to make the active site model. Wire posts imbedded in a suitable base are used to support the amino acid components at the proper elevation.

The active site model may be used to fit substrates and inhibitors and to study stereochemical aspects of its mechanism of action.

IV. INSTRUCTIONS

Program 21 (Coordinate Transformations) transforms x-ray crystallographic coordinates to orthogonal coordinates. It may be used to place a molecule in a particular orientation within a coordinate system. It then provides new coordinates for any set of original coordinates. The original coordinate system may be rectangular (orthogonal) or it may be original x-ray crystallographic data. The program provides for a change of scale, relocation of the origin of the coordinate system, and reorientation of the molecule to a newly defined position. The final coordinates are labeled with the designation of the individual atoms. Distances between any two atoms may also be calculated with this program.

Program 22 (Distances and Angles) provides data regarding the interrelationships of the atoms in a molecule. If program 21 has been used first, program 22 will first transform the original coordinates. It will then provide the distance between two atoms, the angle between three atoms, and the dihedral angle between the two planes defined by the first three and the last three of four consecutive atoms. Provision is made to proceed along a chain of atoms and provide successive dihedral angles as each new atom is added to the chain.

A. Program 21, Coordinate Transformations—Partition 479.59.

Step 1 *Insert* programs 2Ī1, 212.

Step 2. Press E′ to initialize. The memories are cleared and the number 28 is printed.

Step 3. *Enter each parameter* in sequence as it is called for:

28 Scale—1.25 for CPK models, 1.00 if no change.

$\left.\begin{array}{l} 29\ X_f B \\ 30\ Y_f B \\ 31\ Z_f B \end{array}\right\}$ Original coordinates of the atom to be placed at the origin of the new coordinate system. If the orientation of the molecule is not to be determined, press B'. The number 32 will then be printed.

$\left.\begin{array}{l} 32\ a \\ 33\ b \\ 34\ c \\ 35\ \alpha \\ 36\ \beta \\ 37\ \gamma \end{array}\right\}$ X-Ray crystallographic data. If x-ray data are not involved, press A', the number 38 will be printed. If the orientation of the molecule is not to be determined proceed to step 4 when x-ray parameters are entered and 38 is printed.

$\left.\begin{array}{l} 38\ \theta_{XS} \\ 39\ \theta_{YS} \\ 40\ \theta_{ZS} \end{array}\right\}$ These angles define the orientation of the molecule in the new coordinate system. θ_{XS} may have any value. θ_{YS} is $+90°$ or $-90°$. θ_{ZS} is $0°$ or $180°$. A convenient set is $0°$, $90°$, $0°$ for θ_{XS}, θ_{YS}, and θ_{ZS}, respectively.

$\left.\begin{array}{l} 41\ X_f A \\ 42\ Y_f A \\ 43\ Z_f A \end{array}\right\}$ Original coordinates of atom that will lie on the X axis in the new coordinate system.

$\left.\begin{array}{l} 44\ X_f C \\ 45\ Y_f C \\ 46\ Z_f C \end{array}\right\}$ Original coordinates of the third atom that defines the new orientation of the molecule.

Step 4. *Press A* when the number 47 is printed to begin calculations. In about 80 sec the angles θ_Z and θ_Y are printed, and after a short pause, the angle θ_X is printed. These angles are used by the program to define the new orientation of the molecule. The program is still not completed.

Step 5. *Insert programs 213, 214.* Partition 479.59.

Step 6. *Press A'* to complete the calculation. When the calculation is complete (about 80 sec), a number will appear in the display. The coefficients for the three equations that provide the new coordinates are now in memories R_{18}–R_{29}, inclusive. They may be recorded on bank 4 of a new card. They may be recalled and printed.

Step 7. *Press B'* to recall and print the 12 coefficients in order: α_1, α_2, α_3, β_1, β_2, β_3, γ_1, γ_2, γ_3, δ_1, δ_2, δ_3.

Step 8. To *transform* old *coordinates* to new coordinates, enter first the number of the atom, then press A for carbon, B for hydrogen, C for nitrogen, D for oxygen, or E for any other hetero atom. Then enter the original coordinates: X_f, Y_f, and Z_f. When Z_f has been entered, calculation begins automatically and the new coordinates are printed to two decimal places. Program operation returns automatically to the position for the carbon label so that new data may be entered with the label for carbon by simply entering the new atom number and pressing R/S. As before, calculation begins after Z_f has been entered.

Step 9. *Distances* between atoms may be calculated with this program provided the coordinates entered have already been transformed or are originally orthogonal. Press D' and then enter X_A, Y_A, Z_A, and then X_B, Y_B, and Z_B for atoms A and B, respectively. After the last entry, the calculation proceeds automatically. The distance is printed to two decimal places.

B.   Program 22, Distances and Angles—Partition 559.49
(5 0p 17)

Step 1.   *Insert programs* 221, 222, 223.

Step 2.   *Enter coefficients* from prerecorded card; if program 1 was used, the coefficients are already entered. Coefficients may be entered manually and stored in R_{18}–R_{29}, respectively. If coordinates are not to be transformed and are already in an orthogonal coordinate system, this step is not required.

Step 3.   *Press E'* to initialize. The memories are not cleared in this program. New data are simply written over old data as required. Initialization resets the counters.

Step 4.   *Press C if data entered are to be transformed* before use in calculations. If transformation is not required, omit this step. If transformation is required, E' and C must be pressed after each functional calculation is performed (i.e., operations resulting when label A, D, or A' is pressed) but not after entry of the coordinates for each separate atom.

Step 5.   *Enter label* for the first atom. For this program only a single one- or two-digit number can be entered (not a letter).

Step 6.   *Enter the original coordinates* X_f, Y_f, and Z_f for the first atom. If C has been pressed, these are transformed automatically to orthogonal, transformed coordinates X_o, Y_o, and Z_o, and the new coordinates are printed to four decimal places.

Step 7.   *Enter the label and original coordinates for the second atom* as in steps 5 and 6. *Repeat as necessary* until the coordinates for the required number of atoms has been entered: 2 for distance, 3 for an angle, 4 for a dihedral angle.

Step 8.   *Press the appropriate functional key:* D for distance, A for angle, and A' for dihedral angle. The result is printed to four decimal places for a distance or to one decimal place for an angle.

Step 9.   *Press E'* to calculate a new function and begin again with step 3. *However,* if a dihedral angle has been calculated and the next dihedral angle has three consecutive atoms in common with the previous one, it is not necessary to reenter the coordinates of these atoms. *Press B'* (*branch*) if the first three atoms of the new dihedral angle are identical with the first three atoms of the last dihedral angle; *or press C'* (*continue*) if the first three atoms of the new dihedral angle were the last three atoms of the previous dihedral angle. Thus, after the dihedral angle for atoms ABCD has been calculated, press C′ to move the coordinates and prepare for entry of the coordinates of atom E and then calculate the dihedral angle BCDE. If now the dihedral angle BCDF is to be calculated, press B′ before entering the coordinates of the atom F. After the coordinates of E (or of F) have been entered, press A′ to calculate the dihedral angle. Continue with this step as often *as necessary*.

C. Comments

Fixed decimal format may be changed at any time. Simply press GTO XXX, and enter the desired number of decimal places. For program 21, XXX is 013 for coordinates and 424 for distances. For program 22, XXX is 312 and 355 for A′, 394 for C, 526 for D, and 542 for A.

After step 4 of program 21 is completed, the volume of the unit cell is in R_{06}. It may be recalled providing label A′ has not been used. Its value depends, of course, on the scale factor used.

To add or subtract a distance from the transformed coordinates, it is only necessary to place an entry (such as $+ 10 =$, for instance) before the appropriate print command in program 213 at 035, 057, and/or 099. The program is written with blank NOP signals for this purpose.

V.   DESIGN

A.   Introduction

The two programs have a complimentary design. Program 21 provides coefficients when a coordinate transformation is involved and is then used to calculate the locations of atoms and the distances between them. Program 22 can use the coefficients to provide atom locations in a transformed coordinate system. However, it is used mainly to transform orthogonal coordinate data to provide distances, angles, and dihedral angles. Program 21 labels atoms with a number and a letter. Program 22 uses only two-digit numbers to label individual atoms, pairs of atoms for a distance measurement, the three atoms involved in an angle measurement, or the four atoms involved in calculating a dihedral angle.

B.   Cards

Both programs 21 and 22 use two cards. Program 22 requires two cards simply because of its length and the entire program (221, 222, and 223) is entered into the calculator before it is used. (Partition 5 Op 17: 559.49). Program 21 operates in two parts. Part 1 (211 and 212) is on card 1 and is used to enter raw data and begin the calculation of the coefficients. Part 2 (213 and 214) is on card 2. It completes the calculation of coefficients and uses them. Once the calculation of coefficients is complete, they may be recorded on a separate card (bank 4) and used over and over with programs 213 and 214 or program 22.

C. Calculator Labels

Part 1 of program 21 uses only four user-defined labels (Table 2-12). E' is used for initialization; it clears the memories and sets the counters. A' is used to shorten data entry by automatically entering a portion of the data. A begins the calculation of coefficients. B' also shortens data entry.

Part 2 of program 21 uses A, B, C, D, and E as labels for atoms of C, H, N, O, and X, respectively (X may be any atom). A' completes the calculation, B' prints the coefficients, and D' readies program operation for a distance calculation. Calculations begin automatically when data entry is completed with the R/S key.

Label keys for program 22 are defined in Table 2-12. When data entry is complete, the R/S key initiates the calculation.

D. Register Contents

Register contents in program 21 (Table 2-13) are designed to place matrix components in proper position for multiplication. Raw data are

TABLE 2-12

Coordinate Transformations

Label operations		
Program 21		Program 22
Part 1	*Part 2*	
E'—Initialize	A'—Complete	E'—Initialize
A'—Orthogonal	calculation	C—Change (transform
coordinates	B'—Print	coordinates)
B'—Orientation not	coefficients	D—Distance
determined	D'—Calculate	A—Angle
A—Begin calculation	distance	A'—Dihedral Angle
	A,B,C,D,E—Coordinates	B'—Branch (new
	of C,H,N,O,X	dihedral angle)
		C'—Continuation (new
		dihedral angle)

TABLE 2-13

Register Contents

Program 21

	0	1	2	3	4
0	Counter	$0 \to a_3$	$X_1 \to \alpha_3$	$Y_{fB} \to Y_{oB}$	$\theta_{zs} \to \alpha_3$
1	Pgm	$B_1 \to b_1$	$Y_1 \to \beta_1$	$Z_{fB} \to Z_{oB}$	$X_{fA} \to X_{oA} \to \beta_1$
2	$x \to$ Pgm	$B_2 \to b_2$	$Z_1 \to \beta_2$	$a \to A_1$	$Y_{fA} \to Y_{oA} \to \beta_2$
3	3	$0 \to b_3$	β_3	$b \to B_1$	$Z_{fA} \to Z_{oA} \to \beta_3$
4	$3 \to x_i$	$C_1 \to c_1$	γ_1	$c \to B_2$	$X_{fc} \to X_{oc} \to \gamma_1$
5	$t \to y_i$	$C_2 \to c_2$	γ_2	$\alpha \to C_1$	$Y_{fc} \to Y_{oc} \to \gamma_2$
6	Vol $\to z_i$	$C_3 \to c_3$	γ_3	$\beta \to C_2$	$Z_{fc} \to Z_{oc} \to \gamma_3$
7	0	$X_{oi} \to$ Label	δ_1	$\gamma \to C_3$	$\theta_x \to \delta_1$
8	$A_1 \to a_1$	$Y_{oi} \to \alpha_1$	Scale $\to \delta_2$	$\theta_{rs} \to \alpha_1$	$\theta_y \to \delta_2$
9	0 a_2	$Z_{oi} \to \alpha_2$	$X_{fB} \to X_{oB} \to \delta_3$	$\theta_{ys} \to \alpha_2$	$\theta_z \to \delta_3$

Program 22

	0	1	2	3
0	Counter	X_D	α_3	Label
1	X_A	Y_D	β_1	Pgm
2	Y_A	Z_D	β_2	Pgm
3	Z_A	θ_Z	β_3	
4	X_B	θ_Y	γ_1	
5	Y_B	θ_X	γ_2	
6	Z_B	Pointer	γ_3	
7	X_C	Pgm	δ_1	
8	Y_C	α_1	δ_2	
9	Z_C	α_2	δ_3	

entered sequentially with the help of a pointer in R_{00}. R_{03} and R_{04} contain pointers for the matrix operation. Data to be multiplied is placed in R_{07}–R_{16}, inclusive. The multiplier is placed in R_{20}, R_{21}, and R_{22} and the result is extracted from R_{17}, R_{18}, and R_{19}. The final equation coefficients are moved to new locations (R_{38}–R_{49}) so that the calculator partitioning may be changed for the longer program 22. The counter at R_{00} and the pointer in R_{16} assist in the storage and retrieval of data for program 22.

E. Flags

Program 22 uses three flags. Flag 1 is set if the dihedral angle to be calculated is a branch along the chain of atoms from the previous angle. Flag 2 is set to calculate an angle between three atoms. When flag 2 is not set, program operation continues to calculate a dihedral angle. Flag 3 is set if a coordinate transformation using Eqs. (2-1), (2-2), and (2-3) is required.

F. Angle Calculations

The basic calculator transformation involved in all the angle calculations is the polar/rectangular (P/R) key and its inverse (INV P/R). Figure 2-10a,b explains these operations. When the P/R key is to be used, the radius R is placed in the t register and the angle θ in the display. Pressing the P/R keys puts the Y coordinate in the display and the X coordinate in the t register (Fig. 2-10a). The corresponding INV P/R operation is explained in Fig. 2-10b. The diagrams are drawn looking down the Z axis toward the origin so that the angle θ is θ_Z in this instance. Corresponding orientations are involved for θ_X and θ_Y, respectively.

G. Successive Dihedral Angles

When the dihedral angle ABCD (Fig. 2-5) is to be calculated, the required coordinates are arranged sequentially in R_{01}–R_{12}, inclusive (Table 2-11) (see Fig. 2-10c). When the next dihedral angle to be calculated is ABCE (Fig. 2-5), it is only necessary to enter the coordinates of the new atom E since those of AB and C have already been entered. However, in performing the angle calculation, the coordinates of B were subtracted from those of A, C, and D (see Fig. 2-10c) and must be added back to those of A and C before the angle ABCE can be calculated. When this operation is complete, the coordinates of B are subtracted from those of A, C, and E. This may seem to be an awkward maneuver but is most convenient for efficient program operation. A similar maneuver is involved with a different set of atoms when the angle BCEF is calculated after the angle ABCE. This operation is called "continuation" and is illustrated in Figs. 2-5, and 2-10c.

H. Angle Limits

The INV P/R transformation provides angles that lie between $+270°$ and $-90°$. The desired angle lies between $+180°$ and $-180°$. Figure 2-11 explains how this operation is accomplished. Eight possibilities are shown diagrammatically in Fig. 2-11. Three steps are involved. In the first step, $180°$ are subtracted from the angle. The second step depends on the results of step 1. If the result is positive, $180°$ are again subtracted (step 2a) to

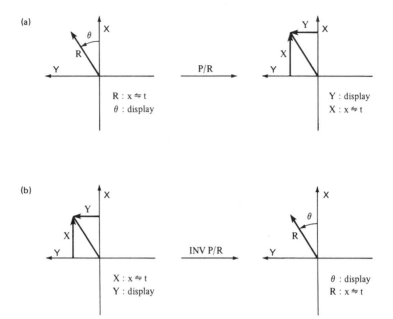

Fig. 2-10 Angle calculations: (a) Polar-to-rectangular transformation; (b) rectangular-to-polar transformation; (c) contents of the storage registers before and after the calculation of a dihedral angle.

give the answer. If the result of step 1 is negative, 360° are added in step 2b. Step 3 depends on the results of step 2b. If step 2b gave a positive angle, then 180° are subtracted to give the answer (step 3a). If, on the other hand, step 2b gave a negative angle, 180° are added to give the answer (step 3b).

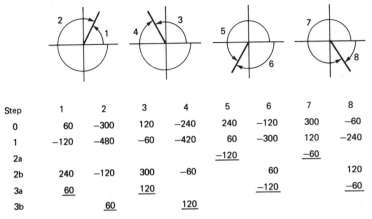

Step	1	2	3	4	5	6	7	8
0	60	−300	120	−240	240	−120	300	−60
1	−120	−480	−60	−420	60	−300	120	−240
2a					−120		−60	
2b	240	−120	300	−60		60		120
3a	60		120			−120		−60
3b		60		120				

Fig. 2-11 Sequence of steps to convert an angle ±360° to an angle ±180°. (1) Subtract 180° from original angle. (2a) If result of step 1 is positive, subtract 180° to obtain final angle. (2b) If result of step 1 is negative, add 360°. (3a) If result of step (2b) is positive, subtract 180° to obtain final angle. (3b) If result of step (2b) is negative, add 180° to obtain final angle. All angles are in degrees and final angles are underlined.

VI. PROGRAMS

A. Program Listings

Program 21 (Coordinate Transformations) requires a partitioning of 479.59 while program 22 (Distances and Angles) is entered with a partitioning of 559.49. The programs are listed in Tables 2-14 to 2-20, inclusive. Label locations follow each program listing. Printer symbol designations and numbers for the notes that follow in Section B are provided in the margins of the tables. The general comments presented in Chapter 1, Section VI, apply here also.

B. Program Notes

1. Programs 211, 212, 213, 214

(1) *Label E'* initializes the calculator by clearing the memories and setting up the counters for the storage of raw data and for the use of the matrix multiplication by the Master Library Module. Subsequently, entry of raw data into the registers R_{28}–R_{46} is accomplished by pressing R/S after each entry into the display. Data storage is accomplished with the

TABLE 2-14

Program 211 Listing

000	76	LBL	(1)	060	32	32	(2)	120	43	RCL	180	22	22
001	10	E'		061	65	×		121	37	37	181	43	RCL
002	47	CMS		062	43	RCL		122	38	SIN	182	31	31
003	03	3		063	33	33		123	95	=	183	42	STO
004	42	STO		064	65	×		124	42	STO	184	23	23
005	03	03		065	43	RCL		125	12	12 B_2	185	36	PGM
006	42	STO		066	34	34		126	43	RCL	186	03	03
007	04	04		067	65	×		127	34	34	187	18	C'
008	02	2		068	53	(		128	65	×	188	43	RCL
009	07	7		069	01	1		129	43	RCL	189	17	17
010	42	STO		070	75	-		130	36	36	190	42	STO
011	00	00		071	43	RCL		131	39	COS	191	29	29
012	69	OP		072	35	35		132	95	=	192	43	RCL
013	20	20		073	39	COS		133	42	STO	193	18	18
014	43	RCL		074	33	X²		134	14	14 C_1	194	42	STO
015	00	00		075	75	-		135	43	RCL	195	30	30
016	99	PRT		076	43	RCL		136	34	34	196	43	RCL
017	91	R/S		077	36	36		137	55	÷	197	19	19
018	99	PRT		078	39	COS		138	43	RCL	198	42	STO
019	72	ST*		079	33	X²		139	37	37	199	31	31
020	00	00		080	75	-		140	38	SIN	200	43	RCL
021	61	GTO		081	43	RCL		141	65	×	201	41	41 (7)
022	00	00		082	37	37		142	53	(	202	42	STO
023	12	12		083	39	COS		143	43	RCL	203	21	21
024	76	LBL		084	33	X²		144	35	35	204	43	RCL
025	16	A'	(2)	085	85	+		145	39	COS	205	42	42
026	01	1		086	02	2		146	75	-	206	42	STO
027	42	STO		087	65	×		147	43	RCL	207	22	22
028	32	32		088	43	RCL		148	37	37	208	43	RCL
029	42	STO		089	35	35		149	39	COS	209	43	43
030	33	33		090	39	COS		150	65	×	210	42	STO
031	42	STO		091	65	×		151	43	RCL	211	23	23
032	34	34		092	43	RCL		152	36	36	212	36	PGM
033	09	9		093	36	36		153	39	COS	213	03	03
034	00	0		094	39	COS		154	54	)	214	18	C'
035	42	STO		095	65	×		155	95	=	215	43	RCL
036	35	35		096	43	RCL		156	42	STO	216	17	17
037	42	STO		097	37	37		157	15	15 C_2	217	42	STO
038	36	36		098	39	COS		158	43	RCL	218	41	41
039	42	STO		099	54	)		159	06	06	219	43	RCL
040	37	37		100	34	√X		160	55	÷	220	18	18
041	03	3		101	95	=		161	43	RCL	221	42	STO
042	07	7		102	42	STO		162	32	32	222	42	42
043	42	STO		103	06	06 V		163	55	÷	223	43	RCL
044	00	00		104	43	RCL		164	43	RCL	224	19	19
045	61	GTO		105	32	32		165	33	33	225	42	STO
046	00	00		106	42	STO		166	55	÷	226	43	43
047	12	12		107	08	08 A_1		167	43	RCL	227	43	RCL
048	76	LBL		108	43	RCL		168	37	37	228	44	44
049	11	A	(3)	109	33	33		169	38	SIN	229	42	STO
050	98	ADV		110	65	×		170	95	=	230	21	21
051	43	RCL		111	43	RCL		171	42	STO	231	43	RCL
052	28	28	(4)	112	37	37		172	16	16 C_3	232	45	45
053	49	PRD		113	39	COS		173	43	RCL	233	42	STO
054	32	32		114	95	=		174	29	29 (6)	234	22	22
055	49	PRD		115	42	STO		175	42	STO	235	43	RCL
056	33	33		116	11	11 B_1		176	21	21	236	46	46
057	49	PRD		117	43	RCL		177	43	RCL	237	42	STO
058	34	34		118	33	33		178	30	30	238	23	23
059	43	RCL		119	65	×		179	42	STO	239	36	PGM

TABLE 2-15

Program 212 Listing

240	03	03		300	71	SBR		360	42	STO		420	43	RCL	
241	18	C'		301	53	(		361	36	36		421	48	48	
242	43	RCL		302	43	RCL		362	43	RCL		422	39	COS	
243	17	17		303	39	39		363	16	16		423	65	×	
244	42	STO		304	95	=		364	42	STO		424	43	RCL	
245	44	44		305	42	STO	θ_Y	365	37	37		425	49	49	
246	43	RCL		306	48	48		366	43	RCL	(13)	426	38	SIN	
247	18	18		307	99	PRT		367	48	48		427	95	=	
248	42	STO		308	43	RCL	(11)	368	39	COS		428	94	+/-	
249	45	45		309	44	44		369	65	×		429	42	STO	
250	43	RCL		310	32	X:T		370	43	RCL		430	11	11	b_1
251	19	19		311	43	RCL		371	49	49		431	43	RCL	
252	42	STO		312	45	45		372	39	COS		432	47	47	
253	46	46		313	22	INV		373	95	=		433	39	COS	
254	43	RCL		314	37	P/R		374	42	STO		434	65	×	
255	29	29	(8)	315	85	+		375	08	08	a_1	435	43	RCL	
256	22	INV		316	43	RCL		376	43	RCL		436	49	49	
257	44	SUM		317	49	49		377	47	47		437	39	COS	
258	41	41		318	95	=		378	39	COS		438	75	-	
259	22	INV		319	37	P/R		379	65	×		439	43	RCL	
260	44	SUM		320	42	STO		380	43	RCL		440	47	47	
261	44	44		321	02	02		381	49	49		441	38	SIN	
262	43	RCL		322	43	RCL		382	38	SIN		442	65	×	
263	30	30		323	46	46		383	85	+		443	43	RCL	
264	22	INV		324	32	X:T		384	43	RCL		444	48	48	
265	44	SUM		325	22	INV		385	47	47		445	38	SIN	
266	42	42		326	37	P/R		386	38	SIN		446	65	×	
267	22	INV		327	85	+		387	65	×		447	43	RCL	
268	44	SUM		328	43	RCL		388	43	RCL		448	49	49	
269	45	45		329	48	48		389	48	48		449	38	SIN	
270	43	RCL		330	95	=		390	38	SIN		450	95	=	
271	31	31		331	37	P/R		391	65	×		451	42	STO	
272	22	INV		332	43	RCL		392	43	RCL		452	12	12	b_2
273	44	SUM		333	02	02		393	49	49		453	43	RCL	
274	43	43		334	32	X:T		394	39	COS		454	47	47	
275	22	INV		335	71	SBR		395	95	=		455	99	PRT	
276	44	SUM		336	53	(		396	42	STO		456	98	ADV	
277	46	46		337	43	RCL		397	09	09	a_2	457	98	ADV	
278	43	RCL		338	38	38		398	43	RCL		458	98	ADV	
279	41	41	(9)	339	95	=		399	47	47		459	91	R/S	
280	32	X:T		340	42	STO	θ_X	400	38	SIN		460	76	LBL	
281	43	RCL		341	47	47		401	65	×		461	53	(	(14)
282	42	42		342	43	RCL		402	43	RCL		462	22	INV	
283	71	SBR		343	08	08	(12)	403	49	49		463	37	P/R	
284	53	(		344	42	STO		404	38	SIN		464	32	X:T	
285	43	RCL		345	32	32		405	75	-		465	42	STO	
286	40	40		346	43	RCL		406	43	RCL		466	05	05	
287	95	=		347	11	11		407	47	47		467	32	X:T	
288	42	STO	θ_Z	348	42	STO		408	39	COS		468	94	+/-	
289	49	49		349	33	33		409	65	×		469	85	+	
290	99	PRT		350	43	RCL		410	43	RCL		470	92	RTN	
291	43	RCL		351	12	12		411	48	48		471	76	LBL	
292	05	05	(10)	352	42	STO		412	38	SIN		472	17	B'	
293	32	X:T		353	34	34		413	65	×		473	03	3	
294	43	RCL		354	43	RCL		414	43	RCL		474	01	1	
295	40	40		355	14	14		415	49	49		475	42	STO	
296	37	P/R		356	42	STO		416	39	COS		476	00	00	
297	43	RCL		357	35	35		417	95	=		477	61	GTO	
298	43	43		358	43	RCL		418	42	STO		478	00	00	
299	32	X:T		359	15	15		419	10	10	a_3	479	12	12	

TABLE 2-16

Program 213 Listing

000	91 R/S	060	95 =	120	43 RCL	180	18 C'
001	42 STO (15)	061	68 NOP	121	48 48	181	43 RCL
002	04 04	062	68 NOP (16)	122	38 SIN	182	17 17
003	99 PRT	063	68 NOP	123	42 STO	183	42 STO β_1
004	91 R/S	064	68 NOP	124	14 14 c_1	184	41 41
005	42 STO	065	99 PRT	125	43 RCL	185	43 RCL
006	05 05	066	43 RCL	126	47 47	186	18 18
007	99 PRT	067	20 20	127	38 SIN	187	42 STO
008	91 R/S	068	65 ×	128	65 ×	188	42 42 β_2
009	42 STO	069	43 RCL	129	43 RCL	189	43 RCL
010	06 06	070	04 04	130	48 48	190	19 19
011	99 PRT	071	85 +	131	39 COS	191	42 STO
012	58 FIX	072	43 RCL	132	95 =	192	43 43 β_3
013	02 02	073	23 23	133	94 +/-	193	43 RCL
014	43 RCL	074	65 ×	134	42 STO	194	35 35 (14)
015	18 18	075	43 RCL	135	15 15 c_2	195	42 STO
016	65 ×	076	05 05	136	43 RCL	196	21 21
017	43 RCL	077	85 +	137	47 47	197	43 RCL
018	04 04	078	43 RCL	138	39 COS	198	36 36
019	85 +	079	26 26	139	65 ×	199	42 STO
020	43 RCL	080	65 ×	140	43 RCL	200	22 22
021	21 21	081	43 RCL	141	48 48	201	43 RCL
022	65 ×	082	06 06	142	39 COS	202	37 37
023	43 RCL	083	85 +	143	95 =	203	42 STO
024	05 05	084	43 RCL	144	42 STO	204	23 23
025	85 +	085	29 29	145	16 16 c_3	205	36 PGM
026	43 RCL	086	95 =	146	43 RCL	206	03 03
027	24 24	087	68 NOP	147	32 32 (19)	207	18 C'
028	65 ×	088	68 NOP (16)	148	42 STO	208	43 RCL
029	43 RCL	089	68 NOP	149	21 21	209	17 17
030	06 06	090	68 NOP	150	00 0	210	42 STO
031	85 +	091	99 PRT	151	42 STO	211	44 44 γ_1
032	43 RCL	092	22 INV	152	22 22	212	43 RCL
033	27 27	093	58 FIX	153	42 STO	213	18 18
034	95 =	094	61 GTO (17)	154	23 23	214	42 STO γ_2
035	68 NOP	095	45 Y×	155	36 PGM	215	45 45
036	68 NOP	096	76 LBL (18)	156	03 03	216	43 RCL
037	68 NOP (16)	097	16 A'	157	18 C'	217	19 19
038	68 NOP	098	43 RCL	158	43 RCL	218	42 STO
039	99 PRT	099	47 47	159	17 17	219	46 46 γ_3
040	43 RCL	100	38 SIN	160.	42 STO	220	43 RCL
041	19 19	101	65 ×	161	38 38 α_1	221	29 29
042	65 ×	102	43 RCL	162	43 RCL	222	94 +/-
043	43 RCL	103	49 49	163	18 18	223	42 STO
044	04 04	104	39 COS	164	42 STO	224	21 21
045	85 +	105	85 +	165	39 39 α_2	225	43 RCL
046	43 RCL	106	43 RCL	166	43 RCL	226	30 30
047	22 22	107	47 47	167	19 19	227	94 +/-
048	65 ×	108	39 COS	168	42 STO	228	42 STO
049	43 RCL	109	65 ×	169	40 40 α_3	229	22 22
050	05 05	110	43 RCL	170	43 RCL	230	43 RCL
051	85 +	111	48 48	171	33 33	231	31 31
052	43 RCL	112	38 SIN	172	42 STO	232	94 +/-
053	25 25	113	65 ×	173	21 21	233	42 STO
054	65 ×	114	43 RCL	174	43 RCL	234	23 23
055	43 RCL	115	49 49	175	34 34	235	36 PGM
056	06 06	116	38 SIN	176	42 STO	236	03 03
057	85 +	117	95 =	177	22 22	237	18 C'
058	43 RCL	118	42 STO	178	36 PGM	238	43 RCL
059	28 28	119	13 13 b_3	179	03 03	239	17 17

TABLE 2-17

Program 214 Listing

240	42	STO	300	98	ADV	360	04	4	420	33	X²
241	47	47	301	68	NOP	361	04	4	421	54	)
242	43	RCL	302	76	LBL	362	69	OP	422	34	ΓX
243	18	18	303	45	Y× (23)	363	04	04	423	58	FIX
244	42	STO	304	91	R/S	364	43	RCL	424	02	02
245	48	48	305	76	LBL (24)	365	17	17	425	99	PRT
246	43	RCL	306	11	A	366	61	GTO	426	98	ADV
247	19	19	307	42	STO	367	33	X²	427	98	ADV
248	42	STO	308	17	17	368	76	LBL (25)	428	61	GTO
249	49	49	309	01	1	369	19	D'	429	19	D' (26)
250	69	OP (20)	310	05	5	370	91	R/S	430	19	D'
251	00	00	311	69	OP	371	42	STO	431	00	0
252	01	1	312	04	04	372	01	01	432	00	0
253	02	2	313	43	RCL	373	99	PRT			
254	42	STO (21)	314	17	17	374	91	R/S			
255	00	00	315	76	LBL	375	42	STO			
256	03	3	316	33	X²	376	02	02			
257	08	8	317	69	OP	377	99	PRT			
258	42	STO	318	06	06	378	91	R/S			
259	01	01	319	81	RST	379	42	STO			
260	01	1	320	76	LBL (24)	380	03	03			
261	08	8	321	12	B	381	99	PRT			
262	42	STO	322	42	STO	382	91	R/S			
263	02	02	323	17	17	383	42	STO			
264	76	LBL	324	02	2	384	04	04			
265	68	NOP	325	03	3	385	99	PRT			
266	73	RC*	326	69	OP	386	91	R/S			
267	01	01	327	04	04	387	42	STO			
268	72	ST*	328	43	RCL	388	05	05			
269	02	02	329	17	17	389	99	PRT			
270	69	OP	330	61	GTO	390	91	R/S			
271	21	21	331	33	X²	391	42	STO			
272	69	OP	332	76	LBL (24)	392	06	06			
273	22	22	333	13	C	393	99	PRT			
274	97	DSZ	334	42	STO	394	53	(			
275	00	00	335	17	17	395	53	(			
276	68	NOP	336	03	3	396	43	RCL			
277	91	R/S	337	01	1	397	01	01			
278	76	LBL (22)	338	69	OP	398	75	-			
279	17	B'	339	04	04	399	43•	RCL			
280	01	1	340	43	RCL	400	04	04			
281	02	2	341	17	17	401	54	)			
282	42	STO	342	61	GTO	402	33	X²			
283	01	01	343	33	X²	403	85	+			
284	01	1	344	76	LBL (24)	404	53	(			
285	07	7	345	14	D	405	43	RCL			
286	42	STO	346	42	STO	406	02	02			
287	00	00	347	17	17	407	75	-			
288	76	LBL	348	03	3	408	43	RCL			
289	85	+	349	02	2	409	05	05			
290	69	OP	350	69	OP	410	54	)			
291	20	20	351	04	04	411	33	X²			
292	73	RC*	352	43	RCL	412	85	+			
293	00	00	353	17	17	413	53	(			
294	99	PRT	354	61	GTO	414	43	RCL			
295	97	DSZ	355	33	X²	415	03	03			
296	01	01	356	76	LBL (24)	416	75	-			
297	85	+	357	15	E	417	43	RCL			
298	98	ADV	358	42	STO	418	06	06			
299	98	ADV	359	17	17	419	54	)			

Program
213, 214
Labels

097	16	A'
265	68	NOP
279	17	B'
289	85	+
303	45	Y×
306	11	A
316	33	X²
321	12	B
333	13	C
345	14	D
357	15	E
369	19	D'

Program
211, 212
Labels

001	10	E'
025	16	A'
049	11	A
461	53	(
472	17	B'

TABLE 2-18

Program 221 Listing

000	91	R/S		060	22	INV		120	42	STO
001	22	INV (1)		061	44	SUM		121	31	31
002	58	FIX		062	02	02		122	43	RCL
003	99	PRT		063	22	INV		123	32	32
004	85	+		064	44	SUM		124	71	SBR
005	43	RCL		065	08	08		125	87	IFF
006	30	30		066	22	INV		126	42	STO
007	65	×		067	44	SUM		127	15	15 θ'_X
008	01	1		068	11	11		128	87	IFF
009	00	0		069	43	RCL		129	02	02 angle
010	00	0		070	06	06		130	89	π
011	95	=		071	22	INV		131	43	RCL
012	42	STO		072	44	SUM		132	10	10 (9)
013	30	30		073	03	03		133	32	X:T
014	91	R/S		074	22	INV		134	43	RCL
015	99	PRT		075	44	SUM		135	11	11
016	72	ST* (2)		076	09	09		136	71	SBR
017	16	16		077	22	INV		137	88	DMS
018	01	1		078	44	SUM		138	42	STO
019	44	SUM		079	12	12		139	17	17
020	16	16		080	76	LBL (7)		140	43	RCL
021	22	INV		081	38	SIN		141	12	12
022	97	DSZ (3)		082	43	RCL		142	71	SBR
023	00	00		083	07	07		143	87	IFF
024	77	GE		084	32	X:T		144	85	+
025	61	GTO		085	43	RCL		145	43	RCL
026	00	00		086	08	08.		146	14	14
027	14	14		087	22	INV		147	95	=
028	76	LBL		088	37	P/R		148	37	P/R
029	77	GE		089	42	STO		149	43	RCL
030	03	3		090	13	13 θ'_Z		150	17	17
031	42	STO		091	43	RCL		151	71	SBR
032	00	00		092	09	09		152	87	IFF
033	98	ADV		093	71	SBR		153	75	-
034	61	GTO		094	87	IFF		154	43	RCL
035	00	00		095	94	+/-		155	15	15
036	00	00		096	85	+		156	95	=
037	76	LBL		097	09	9		157	71	SBR
038	16	A' (4)		098	00	0		158	36	PGM
039	43	RCL		099	95	=		159	76	LBL
040	30	30 label		100	42	STO		160	87	IFF
041	99	PRT		101	14	14 θ'_Y		161	32	X:T
042	87	IFF		102	43	RCL		162	22	INV
043	01	01 branch		103	01	01 (8)		163	37	P/R
044	22	INV (5)		104	32	X:T		164	92	RTN
045	76	LBL		105	43	RCL		165	76	LBL
046	33	X²		106	02	02		166	88	DMS
047	43	RCL		107	71	SBR		167	22	INV
048	04	04 (6)		108	88	DMS		168	37	P/R
049	22	INV		109	42	STO		169	75	-
050	44	SUM		110	32	32		170	43	RCL
051	01	01		111	43	RCL		171	13	13
052	22	INV		112	03	03		172	95	=
053	44	SUM		113	71	SBR		173	37	P/R
054	07	07		114	87	IFF		174	92	RTN
055	22	INV		115	85	+		175	76	LBL
056	44	SUM		116	43	RCL		176	22	INV branch
057	10	10		117	14	14		177	43	RCL
058	43	RCL		118	95	=		178	04	04 (10)
059	05	05		119	37	P/R		179	44	SUM

180	01	01
181	44	SUM
182	07	07
183	43	RCL
184	05	05
185	44	SUM
186	02	02
187	44	SUM
188	08	08
189	43	RCL
190	06	06
191	44	SUM
192	03	03
193	44	SUM
194	09	09
195	61	GTO
196	33	X²
197	76	LBL
198	17	B' (11)
199	86	STF
200	01	01
201	76	LBL
202	30	TAN
203	43	RCL
204	30	30
205	55	÷
206	01	1
207	00	0
208	00	0
209	95	=
210	59	INT
211	95	=
212	42	STO
213	30	30
214	01	1
215	00	0
216	42	STO
217	16	16 (12)
218	87	IFF
219	03	03 change
220	13	C
221	61	GTO
222	00	00
223	00	00
224	76	LBL
225	18	C' (13)
226	22	INV
227	86	STF
228	01	01
229	43	RCL
230	30	30
231	55	÷
232	01	1
233	00	0
234	00	0
235	00	0
236	00	0
237	00	0
238	00	0
239	95	=

TABLE 2-19

Program 222 Listing

240	22	INV	300	09	09	360	10	E'	(16)	420	65	×
241	59	INT	301	01	1	361	00	0		421	43	RCL
242	65	×	302	00	0	362	42	STO		422	15	15
243	01	1	303	42	STO	363	30	30		423	85	+
244	00	0	304	16	16 (12)	364	22	INV		424	43	RCL
245	00	0	305	87	IFF	365	58	FIX		425	27	27
246	00	0	306	03	03	366	03	3		426	95	=
247	00	0	307	13	C	367	42	STO		427	99	PRT
248	00	0	308	81	RST	368	00	00		428	72	ST*
249	00	0	309	76	LBL	369	01	1		429	16	16
250	95	=	310	36	PGM (15)	370	42	STO		430	01	1
251	42	STO	311	58	FIX	371	16	16 (12)		431	44	SUM
252	30	30	312	01	01	372	29	CP		432	16	16 (12)
253	43	RCL	313	29	CP	373	81	RST		433	43	RCL
254	04	04 (14)	314	75	-	374	76	LBL		434	19	19
255	44	SUM	315	01	1	375	13	C (17)		435	65	×
256	07	07	316	08	8	376	22	INV		436	43	RCL
257	44	SUM	317	00	0	377	58	FIX		437	13	13
258	10	10	318	95	=	378	86	STF		438	85	+
259	42	STO	319	77	GE	379	03	03		439	43	RCL
260	01	01	320	34	ГX	380	91	R/S		440	22	22
261	43	RCL	321	85	+	381	99	PRT		441	65	×
262	07	07	322	03	3	382	85	+		442	43	RCL
263	42	STO	323	06	6	383	43	RCL		443	14	14
264	04	04	324	00	0	384	30	30		444	85	+
265	43	RCL	325	95	=	385	65	×		445	43	RCL
266	10	10	326	22	INV	386	01	1		446	25	25
267	42	STO	327	77	GE	387	00	0		447	65	×
268	07	07	328	45	Yˣ	388	00	0		448	43	RCL
269	43	RCL	329	75	-	389	95	=		449	15	15
270	05	05	330	01	1	390	42	STO		450	85	+
271	44	SUM	331	08	8	391	30	30		451	43	RCL
272	08	08	332	00	0	392	91	R/S		452	28	28
273	44	SUM	333	95	=	393	58	FIX		453	95	=
274	11	11	334	99	PRT	394	04	04		454	99	PRT
275	42	STO	335	98	ADV	395	42	STO		455	72	ST*
276	02	02	336	91	R/S	396	13	13		456	16	16
277	43	RCL	337	76	LBL	397	99	PRT		457	01	1
278	08	08	338	34	ГX	398	91	R/S		458	44	SUM
279	42	STO	339	75	-	399	42	STO		459	16	16 (12)
280	05	05	340	01	1	400	14	14		460	43	RCL
281	43	RCL	341	08	8	401	99	PRT		461	20	20
282	11	11	342	00	0	402	91	R/S		462	65	×
283	42	STO	343	95	=	403	42	STO		463	43	RCL
284	08	08	344	99	PRT	404	15	15		464	13	13
285	43	RCL	345	98	ADV	405	99	PRT		465	85	+
286	06	06	346	91	R/S	406	43	RCL		466	43	RCL
287	44	SUM	347	76	LBL	407	18	18		467	23	23
288	09	09	348	45	Yˣ	408	65	×		468	65	×
289	44	SUM	349	85	+	409	43	RCL		469	43	RCL
290	12	12	350	01	1	410	13	13		470	14	14
291	42	STO	351	08	8	411	85	+		471	85	+
292	03	03	352	00	0	412	43	RCL		472	43	RCL
293	43	RCL	353	95	=	413	21	21		473	26	26
294	09	09	354	58	FIX	414	65	×		474	65	×
295	42	STO	355	01	01	415	43	RCL		475	43	RCL
296	06	06	356	99	PRT	416	14	14		476	15	15
297	43	RCL	357	98	ADV	417	85	+		477	85	+
298	12	12	358	91	R/S	418	43	RCL		478	43	RCL
299	42	STO	359	76	LBL	419	24	24		479	29	29

TABLE 2-20

Program 223 Listing

480	95	=		540	95	=	
481	99	PRT		541	58	FIX	
482	72	ST*		542	01	01	
483	16	16		543	99	PRT	
484	01	1		544	98	ADV	
485	44	SUM		545	91	R/S	
486	16	16	(12)	546	76	LBL	
487	61	GTO		547	11	A	(19)
488	13	C		548	86	STF	
489	76	LBL		549	02	02	
490	14	D	(18)	550	61	GTO	
491	43	RCL		551	16	A'	
492	30	30		552	00	0	
493	58	FIX		553	00	0	
494	00	00		554	00	0	
495	99	PRT		555	00	0	
496	53	(					
497	53	(					
498	43	RCL					
499	01	01			**LABELS**		
500	75	-		029	77	GE	
501	43	RCL		038	16	A'	
502	04	04		046	33	X²	
503	54	)		081	38	SIN	
504	33	X²		160	87	IFF	
505	85	+		166	88	DMS	
506	53	(		176	22	INV	
507	43	RCL		198	17	B'	
508	02	02		202	30	TAN	
509	75	-		225	18	C'	
510	43	RCL		310	36	PGM	
511	05	05		338	34	ΓX	
512	54	)		348	45	Yˣ	
513	33	X²		360	10	E'	
514	85	+		375	13	C	
515	53	(		490	14	D	
516	43	RCL		531	89	π'	
517	03	03		547	11	A	
518	75	-					
519	43	RCL					
520	06	06					
521	54	)					
522	33	X²					
523	54	)					
524	34	ΓX					
525	58	FIX					
526	02	02					
527	99	PRT					
528	98	ADV					
529	91	R/S					
530	76	LBL					
531	89	π'	angle				
532	43	RCL					
533	31	31					
534	22	INV					
535	37	P/R					
536	94	+/-					
537	85	+					
538	09	9					
539	00	0					

indirect address. If the raw data are already expressed as rectangular coordinates, then A' is pressed to place unity in place of the x-ray crystal unit cell dimensions, a, b, and c and 90° for α, β, and γ, the angles between the axes of the unit crystal.

(2) *Label A'* is used when x-ray crystallographic data are not involved. It automatically places 1 in R_{32}, R_{33}, and R_{34} and 90 in R_{35}, R_{36}, and R_{37}. It resets the counter at R_{00} so that the remaining data are entered into the proper registers.

(3) *Label A* begins the calculation of the coefficients for the three transformation equations, (2-1), (2-2), and (2-3) (see Section II).

(4) The new scale in R_{28} is applied to the x-ray cell dimensions in R_{32}, R_{33}, and R_{34}.

(5) The matrix components for the transformation of x-ray coordinate data to rectangular coordinates are computed and placed in position according to Eq. (2-4) (see Section II). The volume of the unit cell is calculated first and stored in R_{06}. It may be recalled later, if desired, for comparison with the value reported by the original investigators. Next, the values of A_1, B_1, B_2, C_1, C_2, and C_3 are calculated and placed in the appropriate positions for matrix multiplication.

(6) The sequence of steps 174–200 places the fractional coordinates of the atom B in position for matrix multiplication, carries out the multiplication with the help of the Master Library Module, and then returns the resulting rectangular coordinates to the locations of the original coordinates.

(7) The process of step 6 is repeated for the coordinates of the atom A (201–227) and atom C (228–254).

(8) The coordinates of the atom B are now subtracted from those of A and C.

(9) The angle θ_Z is now calculated. This is the angle through which the original rectangular coordinate system must rotate to reach its new position. $\theta_Z = \theta_{ZS} - \theta'_Z$, where θ_{ZS} is the selected standard angle about the Z axis (see Section II), and θ'_Z is the original angle that BA makes with the BX axis in the $X-Y$ plane (see Fig. 2-4a). The angle θ'_Z is obtained with the rectangular-to-polar conversion feature. Since this transformation is used several times, it is provided with a subroutine, LBL ((step 461). The resulting angle θ_Z is printed. See Fig. 2-10 for an explanation of the P/R and INV P/R transformations.

(10) The angle θ_Y about the Y axis that the coordinate system must move to place A on the X axis is now calculated in a manner similar to θ_Z (see Fig. 2-4b). The result is printed.

(11) To find θ_X, the Y and Z coordinates of the atom C, which result when the coordinate system is moved through the angles θ_Z and θ_Y,

are first calculated. Then the angle θ'_X is calculated (see Fig. 2-4c) and, finally, θ_X.

(12)   The coefficients A_1, B_1, B_2, C_1, C_2, and C_3 are now moved to new locations, R_{32}–R_{37}, so that the matrix registers may be used again.

(13)   The coefficients a_1, a_2, a_3, b_1, and b_2 of Eq. (2-5) are now calculated and placed in position for matrix multiplication.

(14)   Label (performs the rectangular-to-polar conversion for calculation of rotation angles. See Note 9.

(15)   The sequence 001–091 stores x-ray coordinates and then transforms them to rectangular coordinates according to Eqs. (2-1), (2-2), and (2-3).

(16)   A sequence $+XX=$ may be added to change the final location of the molecule.

(17)   LBL y^x transfers operation to the label for a carbon atom (step 303).

(18)   Label A′ completes the calculation of the coefficients of Eq. (2-5), namely b_3, c_1, c_2, and c_3, and places them in position for matrix multiplication.

(19)   The final matrix multiplications provide the coefficients of Eqs. (2-1), (2-2), and (2-3). The α coefficients are obtained first with A_1 in place of X_R in Eq. (2-5) and 0 in place of Y_R and Z_R. Then the β coefficients are obtained with B_1 in place of X_R, B_2 in place of Y_R, and 0 in place of Z_R. For the γ coefficients, C_1 is in place of X_R, C_2 in place of Y_R, and C_3 in place of Z_R. Finally, D_1 $(-X_R$ of B) is used in place of X_R, D_2 $(-Y_R$ of B) is used in place of Y_R, and D_3 $(-Z_R$ of B) is used in place of Z_R. This multiplication provides the δ terms of Eqs. (2-1), (2-2), and (2-3).

(20)   The print registers are cleared so that labels A, B, C, D, and E may be used.

(21)   The steps 252–275 set up new pointers and then move the coefficients to new locations (see Table 2-14) so that they may be recorded on bank 4 of a card and so that program 22, which uses program registers of bank 3, will not obliterate these data. LBL NOP is used for the repetitive sequence.

(22)   *Label B′* sets up pointers and prints the coefficients.

(23)   See Note 17.

(24)   *Labels A, B, C, D, and E* set up the print registers so that each succeeding calculation is preceded by an appropriate label for the individual atom. The program in each case goes to the beginning for storage of the raw data before the calculation of the coordinate transformation by step 014. It is possible with this program to continue each transformation without inserting the identifying label.

(25)   *Label D′* permits distance calculations provided the input is rec-

tangular coordinate data. The coordinates are first stored and then used in the calculation following entry of the last coordinate.

(26) Operation is returned to D' to permit further distance calculations. Paper advances separate data in the printout.

2. *Programs 221, 222, 223*

(1) *Invert fix* allows the atom label to be printed in a shorter format than its coordinates. This sets the coordinate data apart and improves the appearance of the output. R_{30} is used to store the atom labels. The factor of 100 moves the label to the left two digits in preparation for the label of the next atom.

(2) R_{16} is the pointer for storage of raw data.

(3) This sequence sets each coordinate set apart with a paper advance.

(4) *Label A'* (dihedral angle) is pressed following the entry of four coordinate sets to calculate the rotation angle.

(5) Flag 1 is set if the angle to be calculated is a branch and label B' was pressed (step 198).

(6) This sequence sets the atom B at the origin of the coordinate system by subtracting its coordinates from those of A, C, and D.

(7) Calculation is illustrated in Fig. 2-4 (A and C reversed). The angle θ_Z' required to move C to the X–Z plane is first calculated using the INV P/R transformation (see Fig. 2-10). θ_Z' is stored in R_{13}. R (now X_c) is in the t register. The Z coordinate is recalled from R_{09} and after INV P/R the display contains θ_Y' (Fig. 2-4b). This angle is subtracted from 90° to give θ_Y, which is stored in R_{14}.

(8) Calculation of θ_X is now carried out (Fig. 2-4a, A and C reversed). The bond BA must first be moved through the angles θ_Z and θ_Y. These movements provide values of Z_A and Y_A from which the angle θ_X is obtained. The rotation about the Y axis produces X_A, which is stored in R_{20}. The final rotation about the X axis leaves Y_A in the t register. If the angle ABC is to be calculated, flag 2 is set and program operation branches to label π (step 531). Label π calculates the angle θ_Z' (Fig. 2-3a). When this angle is added to 90°, the angle ABC is obtained.

(9) If flag 2 has not been set, program operation continues to calculate the rotation angle ABCD (θ_X' in Fig. 2-3a). The coordinates of D must first be transformed by moving the bond CD through the angles θ_Z, θ_Y, and θ_X in turn. Finally, Z_D is placed in the display and Y_D in the t register. The operation INV P/R now gives θ_X', which is the rotation angle ABCD. Program operation then moves to label PGM at step 310 (note 15).

(10) When a branch occurs, the coordinates of the rotation angle

ABCD are in the registers R_1–R_{12}; those of A being in R_1–R_3, those of B in R_4–R_6, and those of C in R_7–R_9. However, the coordinates of B have been subtracted from those of A and C (see note 4). LBL INV restores the coordinates of A and C to their original values by adding back the coordinates of B (see Fig. 2-10).

(11) *Label B'* is pressed following the operation of A', if the new angle ABCE has its first three atoms identical with the first three atoms of the previous angle ABCD. After B' is pressed, data entry of the label and coordinates of atom E is followed by pressing label A' to calculate the angle. Flag 1 is set for a branch. Pressing B' also must change the atom labels. The use of the integer key is a convenient way to lose the unwanted label. See also how the label is changed for C' at step 240.

(12) R_{16} is the pointer for storage of new coordinates. It is reset when a major operation is completed.

(13) *Label C'—Continue—*is pressed following the operation of A' if the new angle, BCDE has as its first three atoms the last three atoms of the previous angle ABCD. After C' is pressed, data entry of the label and coordinates of E is followed by pressing A' to calculate the angle.

(14) Pressing C' indicates that the angle BCDE to be calculated follows that of ABCD so that the first three atoms of the new angle are the same as the last three atoms of the first angle. The coordinate change is accomplished with the sequence beginning at step 253. Note that this sequence first adds the coordinates of B to those of C and D and then moves all the coordinates down to their new locations (see Fig. 2-10).

(15) Label PGM is a sequence that converts an angle that may lie between $-360°$ to $+360°$ as a result of a series of INV P/R transformations, to one which lies between $-180°$ and $+180°$ (see Fig. 2-11 for an explanation). Figure 2-3e defines a positive angle. In Fig. 2-3a the bond DC would have to move clockwise to bring it to the X–Y plane.

(16) *Label E'—Initial—*is pressed to begin a calculation. It clears the label register R_{30}, removes fix decimal format, clears the t register, sets up the counters, and sends program operation to the beginning—an operation which resets the flags. Data entry is followed by pressing R/S after each item. The first item is a one- or two-digit number label for the atom. It must be entered so that the next entry of a coordinate may be processed correctly. The raw coordinate data are entered next, X, Y, and Z in that order. After each coordinate set is entered, the paper advances or, if the coordinates are to be changed, the new coordinates are printed. In this case, the paper does not advance because the label or function calculation sets the coordinate apart in an attractive format.

(17) *Label C—Coordinate Change—*is pressed after E' if the coordinates are to be transformed. Unless E' is pressed, all subsequent raw data

entries will automatically result in the transformed coordinates being calculated and printed. Flag 3 is set for a coordinate change.

(18) *Label D—Distance*—is pressed after two sets of coordinates have been entered to calculate the distance between the two atoms.

(19) *Label A—Angle*—is pressed after three sets of coordinates have been entered to calculate the angle between them. Flag 2 is set for an angle calculation.

REFERENCES

1. P. Gund, J. D. Androse, J. B. Rhodes, and G. M. Smith, Three-dimensional molecular modeling and drug design. *Science* **208**, 1425, 1980.
2. C. L. Strong, The amateur/Scientist: How an amateur can construct a model of an enzyme model at modest cost. *Sci. Am.* **234**, 124, 1976.
3. F. H. Clarke, New skeletal space-filling models. *J. Chem. Ed.* **54**, 230, 1977.
4. F. H. Clarke, H. Jaggi, and R. A. Lovell, Conformation of 2,9-dimethyl-3'-hydroxy-5-phenyl-6,7-benzomorphan and its relation to other analgetics and enkephalin. *J. Med. Chem.* **21**, 600, 1978.
5a. W. B. Lawson and G. J. S. Rao, Specificity in the alkylation of methionine at the active site of α-chymotrypsin by α-bromoamides. *Biochemistry* **19**, 2133, 1980.
5b. W. B. Lawson, Specificity in the alkylation of serine at the active site of α-chymotrypsin by aromatic α-bromoamides. *Biochemistry* **19**, 2140, 1980.
6a. S. R. Wilson and J. C. Haffman, Cambridge data file in organic chemistry. Applications to transition-state structure, conformational analysis, and structure/activity studies. *J. Org. Chem.* **45**, 560, 1980.
6b. J. P. Tollenaere, H. Moereels, and L. A. Raymaekers, "Atlas of the Three-Dimensional Structure of Drugs." Elsevier/North Holland Biomedical Press, New York, 1979.
7. S. C. Nyburg, Some uses of a best molecular fit routine. *Acta Crystallogr.* B30, 251, 1974.
8. W. R. Kester and B. M. Matthews, Comparison of the structure of carboxypeptidase A and thermolysin, *J. Biol. Chem.* **252**, 7704, 1977.
9. R. M. Garavito, M. G. Kossmann, P. Argos, and W. Eventoff, Convergence of active center geometries. *Biochemistry* **16**, 5065, 1977.
10. H. Goldstein, "Classical Mechanics," p. 107. Addison-Wesley, Reading, Massachusetts, 1950.
11. "Computing Methods in Crystallography" (J. S. Rollet, ed.), p. 22. Pergamon Press, New York, 1965.
12. R. S. Ledley, "Use of Computers in Biology and Medicine," Chapter 16. McGraw-Hill, New York, 1965.
13. S. R. Wilson and J. C. Huffman, The structural relationship of phorbol and cortisol: A possible mechanism for the tumor-promoting activity of phorbol. *Experimentia* **32**, 1489, 1976.
14. P. J. Roberts, J. C. Coppola, N. W. Isaacs, and O. Kennard, Crystal and molecular structure of cortisol ($11\beta,17\alpha,21$-trihydroxy-pregn-4-ene-3,20-dione)methanol solvate. *J. Chem. Soc.* **II**, 774, 1973.
15. J. P. Leclercq, G. Germain, and M. Van Meerssche, 17α-21-Dihydroxy-4-pregnene-3,11,20-trione, cortisone. *Crystallogr. Struct. Commun.* **1**, 13, 1972.

16. R. C. Petterson, G. I. Birnbaum, G. Ferguson, K. M. S. Islam, and J. G. Sime, X-Ray investigation of several phorbol derivatives. The crystal and molecular structure of phorbol bromofuroate–chloroform solvate at $-160°$. *J. Chem. Soc. B*, 980, 1968.

17. D. G. Morris, P. Murray-Rust, and J. Murray-Rust, A conformational study of bicyclo[3.1.0]hexane—crystal and molecular structure of N'-isopropylidenebicyclo-[3.1.0]hexane-6-exo-carbohydrazide. *J. Chem. Soc.* II, 1577, 1977.

18. P. D. Chadwick, Crystal structure of the growth inhibitor, maleic hydrazide (1,2-dihydropyridazine-3,6-dione. *J. Chem. Soc.* II, 1386, 1976.

19. J. P. Glusker, H. L. Carrell, H. M. Berman, B. Gollen, and R. M. Peck, Alkylation of a tripeptide by a carcinogen: The crystal structures of sarcosylglycine, 9-methyl-10-chloromethylanthracine and their reaction product, *J. Am. Chem. Soc.* **99**, 595, 1977.

20. F. C. Bernstein, T. F. Koetzle, G. T. B. Williams, E. F. Meyer, Jr., M. D. Price, J. R. Rogers, O. Kennard, T. Shimanouchi, and M. Tasumi, *J. Mol. Biol.* **112**, 535, 1977.

21. LABQUIP, Ashridgewood Place, Wokingham RG11 5RA, England.

CHAPTER 3

Potentiometric Titrations

I. INTRODUCTION

A. General Considerations

At first glance, the solution of the ionization equation $K = [H^+][A^-]/[HA]$ would not seem to offer a test for the capacity of a programmable calculator. However, when the equation is applied to the real situations that permeate chemistry and biology, formidable challenges are provided for the calculator and for the computer. It is anticipated that an understanding of the principles described in this chapter will equip the reader to tackle other problems of chemical equilibria and possibly of enzyme kinetics.

Nowhere is the link between chemistry and biology more clearly illustrated than in the study of chemical ionization in solution. Our study will begin with the purely chemical phenomenon of acid–base dissociation and will lead us to its application to the partitioning of organic compounds between polar and nonpolar solvents. In the next chapter we shall see how the partition coefficient can be applied to help explain the biological action of chemicals in living systems.

The capacity of the Texas Instruments programmable calculator (TI-59) to solve chemical problems is well illustrated in its application to potentiometric titrations of weak acids and bases. The programs described in this chapter involve the storage and retrieval of raw data, the repeated performance of cumbersome calculations, and the application of iterative operations to solve complex equations.

The nature of the tasks performed by the calculator will be discussed first, and then illustrated with worked examples. This will be followed by instructions for using the programs, the programs themselves, and finally a detailed analysis of the programs and how they operate. It is anticipated that many readers will find the programs very useful in their present form. For them, the examples, the instructions, and the program listings are most important. Other readers will want to know how the programs are constructed so that they may be altered to solve a different problem or adapted to serve a particular need. For these readers, the program notes, the memory contents, and the decision maps are very important. To assist these needs, each aspect is treated in turn for all the programs.

For those interested in quantitative drug design, one of the most useful constants is the partition coefficient (P or log P), which describes the distribution of a drug between lipids and water [1–4]. The determination of these constants usually involves the measurement of concentrations, often a time-consuming process. However, it has been found that P can be simply determined for an ionizable substance from the results of potentiometric titrations (see below).

Ionization constants are used to describe the strengths of acids and bases [5,6]. They reveal the proportions of the different ionic species that result when acids, bases, and their salts are dissolved in water.

Potentiometric titration is one of the most useful methods to determine an ionization constant K_a or its negative logarithm, the pK_a. During such a determination an S-shaped titration curve (Fig. 3-1) is usually obtained in which the pH of the solution is plotted against the volume of titrant (strong acid or strong base) added. The titration curve usually has a typical "break" or "end point," which results when small incremental additions of titrant (X) cause comparatively large changes in the pH of the solution. When this portion of the curve is symmetrical, the end point may be determined graphically using a set of equidistant parallel lines (Fig. 3-2). The titrant volume at this end point is known as the equivalent volume

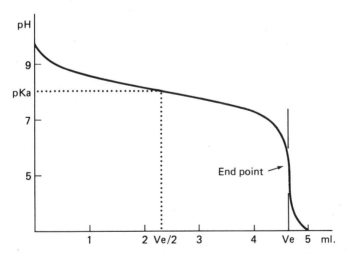

Fig. 3-1 Titration curve: A strong base, "tris," is titrated with $0.1N$ HCl (Section II.B). Values of pH are plotted against titrant volume. The end point is reached when the slope of the curve changes sign. The equivalent volume (V_e) is the volume of titrant at the end point. The pK_a is the pH at one-half the equivalent volume.

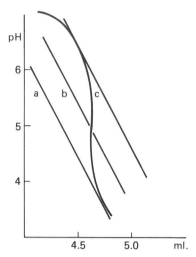

Fig. 3-2 End point by parallel lines: Part of the curve of Fig. 3-1 is drawn on a larger scale. Parallel lines a and c are tangent to the titration curve. Parallel line b is equidistant from lines a and c and parallel to them. The end point is the intersection of line b with the titration curve.

(V_e) (Fig. 3-1). From V_e the equivalent weight (W_e) of the compound titrated is easily calculated. For compounds with a pK_a near 7.0, the pH at one-half V_e (Fig. 3-1) is the pK_a of the compound. For stronger acids or stronger bases, a correction must be applied to account for the contribution to the pH of the dissociation of water. As illustrated in Fig. 3-1, the graphical method provides only a single value of the V_e and pK_a, respectively, even though a series of many single measurements may have been involved in the preparation of the titration curve.

Albert and Serjeant [5] have described in detail how to use the series of pH measurements obtained during a potentiometric titration to indicate the precision of the pK_a determination. They considered the equivalent weight to be a known constant for the compound titrated. In chemical research, this is not always the case; in fact, the determination of the equivalent weight may be one of the most useful results of a potentiometric titration. In other instances, the amount of compound present in the solution may not be known precisely. For these reasons, *programs 31 and 32* were designed to locate both the equivalent volume (V_e) and the pK_a. When the sample weight is known, this method also provides the equivalent weight (W_e).

Kolthoff and Furman [7] have provided a useful mathematical method for locating the end point of a potentiometric titration. The method has been carefully analyzed by Fortuin [8] and more recently by Ebel and

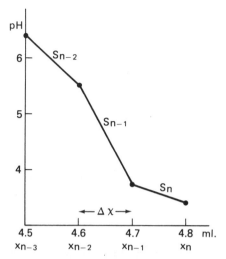

Fig. 3-3 End point by the Kolthoff method: Slopes (S) are calculated between successive titration points that are equidistant apart. The four points surrounding the maximum slope are selected. Equation (3-10) then provides the equivalent volume (V_e). See Section III.B.

Seuring [9]. The method involves the differences between the slopes of the titration curve at successive intervals near the end point (Fig. 3-3). *Program 31* pK_a uses this convenient method to advantage and provides the V_e, pK_a, and a measure of the precision of the pK_a determination.

For dilute solutions ($0.002\,M$) of compounds with a pK_a of less than 4.5 or greater than 9.5, one end of the titration curve is very nearly a straight line with no detectable break or inflection. The other end of the titration curve is usually well defined. Since the titration curves of acids and bases with the same pK_a are mirror images, the defined end can be treated as the beginning for purposes of calculation (see Fig. 3-4 and also Fig. 3-7 in Section VI.A). For titration curves such as these, the changes in slope near the end point are so small that the Kolthoff method fails. In fact, automatic derivative methods also fail under these circumstances. Fortunately, Barry and Meites [10] have shown that computer techniques can be used in such cases to calculate both the equivalent volume and the pK_a from a set of titration data. Their method minimizes the difference between calculated and observed pH values at each titration point. Briggs and Stuehr [11] have used the constancy of the pK_a values to locate the equivalence point. In a comment on this method, Meites *et al.* [12] note that while computer time is saved in this way, the results may not be as precise as those obtained using the method of Barry and Meites [10]. We have found that the Briggs and Stuehr method can be easily adapted for

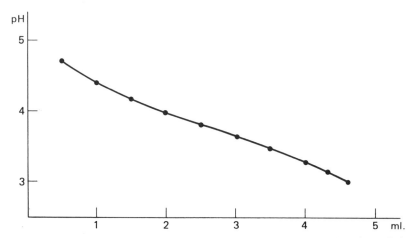

Fig. 3-4. Titration curve of a weak base with a strong acid: Although there is no obvious end point to this curve, the iterative mathematical method of program 32 provides the equivalent volume and the pK_a. See Section III.C.

programming on the TI-59 calculator. The results are truly amazing to someone who has depended on the graphical method for end-point determination. *Program 32* is based on the observation that the standard deviation (SD) of the pK_a values calculated for each experimental point in an acid–base titration curve is a minimum when the equivalent volume (V_e) has the theoretical value.

Programs 32 pK_a and equivalent volume use a number of observations to determine both V_e and pK_a by means of an iterative procedure. An initial value (V'_e) is selected and the SD of the pK_a values is obtained. V_e is then varied systematically until a minimum SD is obtained for the pK_a values. The data are also used to calculate 95% confidence limits for both V_e and the mean pK_a. Since the equations are nonlinear, the calculation of the 95% confidence limits is not simple. Our approach (see Section II) is based on those of other authors [12–15] and provides *approximate* limits.

When the end point of an aqueous titration is not distinct, it is often an advantage to repeat the titration in the presence of octanol as described in the example in Section III.D. In this case, the titration curve is moved in the direction of pH 7.0, and there is usually a distinct end point. The calculation of the apparent pK_a (pK') under these circumstances is straightforward (see Section II), and the results not only provide the V_e for the aqueous titration but are used to calculate the partition coefficient (P) as well. This situation is illustrated in Fig. 3-5 of Example D for the case in which the un-ionized species precipitates during the titration. It is usually an advantage to add an excess of the salt-forming acid or base before

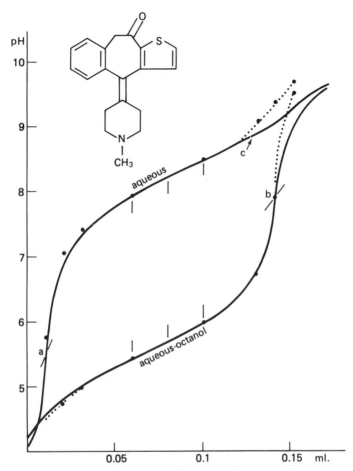

Fig. 3-5 Titration curves to provide the partition coefficient: An organic base (see structure) is titrated in water and in a mixture of water and octanol. The initial volume *a* and the equivalent volume *b* are determined with parallel lines. Program 33 is used to calculate three pK_a values (at the positions indicated by the short vertical lines) for each curve and then to calculate log *P*. A precipitate forms at the inflection point c. Equation (3-19a) may be used to calculate the solubility using the pH and titrant volume at c. The solid dots and the dotted curves are calculated using program 34. See Sections III.D and III.E.

titrating with strong base or acid, respectively. This technique assures an S-shaped curve to define more clearly the beginning of the titration.

The use of potentiometric titrations to determine log *P* was described independently in two laboratories. Kaufman *et al.* [16] calculated log *P* from the difference between the pK_a measured in the absence of octanol

and the apparent acid dissociation constant (pK') measured when octanol
was present during the titration. Seiler [17] used a dilute aqueous solution
in which the titration was carried out half-way to determine the pK_a and
then a small volume of octanol was added and the titration was completed
to determine pK'. From these constants, the partition coefficient was
calculated.

Martin [18] has discussed the potentiometric titration method of determining log P and concludes that it may have limitations. Among the
limitations are the need for constant temperature during the titration. This
is an important consideration for any method since the partition coefficient
is a function of temperature. Seiler used octanol-saturated water for his
aqueous titration and found that the octanol tended to come out of solution.

We have found it unnecessary to saturate mutually the solvents before
the experiment—the volume changes from those measured appear negli-
gible. The most dramatic finding in our experiments is the rapid equilibra-
tion of the solvent mixture and the clear demonstration that the pH values
need no correction, but can be used directly for calculations. When the
volume ratios are of the order of 5–20 ml water and 5–20 ml of octanol,
mixing is not a problem and the pH meter readings vary smoothly as the
titrant is slowly added. Figure 3-5 shows a set of titration curves that may
be used for the calculation of the log P.

The potentiometric titration method has an added advantage in allow-
ing the immediate calculation of the apparent partition coefficient [19] (P'
and log P') at any desired pH such as the physiological pH of 7.4 [20].
References [1–4] provide extensive discussions of the usefulness of log P
and log P' for the study of quantitative structure–activity relationships
(QSAR). The value of log P' for QSAR studies has been emphasized by
Scherer and Howard [20], who use the term distribution coefficient for this
constant.

Program 33, partition coefficients, has been written to calculate the pK_a
and pK' values directly from laboratory potentiometric titrations and to
convert these to P, log P, P', and log P' values. The program has prompting messages for entering the constants and the results are labeled so that
the printout can be incorporated directly in the laboratory notebook as a
record of the determination.

Program 34, titration curve, has also been written for the construction
of pH titration curves from the physical constants: pK_a, P, N, and the
volumes of solvents and titrant. This program is especially useful to confirm the shape of an experimental titration curve with or without the
presence of octanol or to detect alterations in its shape that could indicate
an impurity or an additional titratable group. Program 34 considers only

monoprotic acids and bases. Other authors have described calculator programs for monoprotic [21] and polyprotic [22] acids and bases, without considering the partition coefficient. A different approach, termed the master variable concept, has been described for use with programmable calculators [23], and a program has been written in which the volume of titrant is calculated exactly as a function of the hydrogen ion concentration [24].

B. Summary of Programs

Program 31 was written to locate rapidly the end point and to calculate the pK_a when the titration of an acid, base, or salt is carried out using a burette and a pH meter. The operation of program 1 is rapid and applies when the titration curve (pH of solution versus volume of titrant) has a well-defined break. The slopes of the curve between successive points are plotted by the printer. A peak in the graph indicates that the end point is passed. The end point is then calculated from the slopes using the Kolthoff method. The results are printed: invidivual pK_a values for each point, equivalent volume, pK_a, and standard deviation (SD).

Program 32 is used when there is no apparent break in the titration curve or when the end point cannot be located precisely either from data generated manually or read from the curve generated by an automatic titrator. The program has been written to vary V_e systematically, and by iteration to determine the value of V_e that provides the minimum SD for the pK_as. As explained above, SDs are calculated for both the V_e and the mean pK_a ($\bar{p}K_a$) using nonlinear regression analysis. These are transformed into 95% confidence limits for the corresponding equivalent weight and $\bar{p}K_a$. The program determines the proper number of degrees of freedom (ν) $(n - 2)$, and the corresponding Student's t value is calculated and then used by the program.

The printout consists first of the individual pK_a values calculated with the initial guess V_e' of the equivalent volume. These enable the operator to check the data for inconsistency. Next, each SD is printed so the operator can observe that convergence to an answer will occur. Printing of each SD is preceded by a flashing display of the V_e and pK_a. Finally, the resultant pKs are printed for each point, followed by the V_e, its 95% confidence interval (CI), the mean pK_a and its 95% CI. The 95% CI is always larger than SD, and depends on the number of data points used.

Program 33 enables one to use titration data to calculate the partition coefficient (P) and its logarithm $(\log P)$. The titration is carried out twice, first in water and then in water with octanol present. The latter titration provides an apparent acidity constant (pK') and it is the difference between pK' and pK_a that is used to calculate P. Program 33 also enables one to calculate the apparent partition coefficient ($\log P'$) at any desirable pH (such as the physiological pH of 7.4). Often it is not possible to perform a complete titration in water alone because the un-ionized compound is not sufficiently soluble. In this case, a portion of the titration curve is often sufficient to provide the pK_a. *Program 33* has been written to handle a number of situations. It is versatile and is applied for acid or base titrations of weak bases or acids, respectively, or of their salts. The program can be used in conjunction with an automatic titrator or with a burette and pH meter. Both pK_a and pK' may be calculated followed by the P constants: P, $\log P$, P' and $\log P'$. If pK_a is known, the program calculates only pK' and then the P constants. When both pK and pK' are known, the program calculates the P constants. On the other hand, the program can be used simply to calculate individual pK_a values for a number of points on a titration curve. When an automatic titrator is used, the formation of a precipitate may result in an inflection in the titration curve. The pH and titrant volume at this inflection point can be used to calculate the water solubility of the precipitate.

Program 34 provides the data needed to construct a titration curve for the titration of a weak acid, a weak base, or the salt of one of these with a strong acid or base in the presence or absence of octanol. The program asks for the required constants and prints the result when the calculation is complete. Two situations are considered for successive points, either the volume of octanol, the initial volume of water, and the volume of titrant are reentered, or only the titrant volume is reentered. The reentry of titrant volumes can be made automatic as described in Section VIII.B.

II.   CALCULATIONS

A.   General Equations

General equations will be derived for the titration of weak monoprotic acids and bases and their salts. It will be assumed that the temperature is constant during the titration, that all equilibria are rapidly attained, and that ions are not soluble in octanol. Experience has shown that these conditions are easily met for many compounds. In fact, it has been found that titration curves generated experimentally in the presence of octanol are reproduced almost precisely by calculator techniques that solve the equations derived below. The following symbols are involved. All concentrations [ ] are in moles per liter.

BH^+	cation	K_a	acid dissociation constant
B	weak base	K'	apparent dissociation constant
A^-	anion		(in presence of octanol)
HA	weak acid	K_w	water dissociation constant
P	partition coefficient	V_e	equivalent volume of titrant
	(neutral molecule)	V_i	initial volume of aqueous
P'	partition coefficient		solution
	(at given pH)	V_w	total volume of aqueous solution
f	partition factor	V_o	volume of octanol
[ ]	concentration term	X	volume of titrant
o	subscript for octanol	N	normality of titrant
w	subscript for water		

Four cases are considered:

1. Titration with a strong acid:
    (1) $B + H^+Cl^- \rightarrow BH^+ + Cl^-$
    (2) $Na^+A^- + H^+Cl^- \rightarrow HA + Na^+ + Cl^-$
2. Titration with a strong base:
    (3) $HA + Na^+OH^- \rightarrow A^- + Na^+ + H_2O$
    (4) $BH^+Cl^- + Na^+OH^- \rightarrow B + Na^+ + Cl^- + H_2O$

The following relationships are general:

$$pH = \log \frac{1}{[H^+]}; \quad pK_a = \log \frac{1}{K_a}; \quad pK' = \log \frac{1}{K'}$$

$$f = \frac{V_oP}{V_w} + 1; \quad V_w = V_i + X; \quad [X] = \frac{XN}{V_w}$$

$$[V_e] = \frac{V_eN}{V_w}$$

For titrations with a strong acid: $[X] = [Cl^-]$.
For titrations with a strong base: $[X] = [Na^+]$.
For all titrations, V_e equals the sum of the amounts of all components of the titrated compound.
The following equilibria are involved:

$$H_2O \overset{K_w}{\rightleftarrows} H^+ + OH^-; \quad K_w = [H^+][OH^-]$$

For weak acids and their salts:

$$HA \overset{K_a}{\rightleftarrows} H^+ + A^-; \quad K_a = [H^+][A^-]/[HA]$$

In the presence of octanol:

$$HA_w \overset{P}{\rightleftarrows} HA_o; \quad P = [HA_o]/[HA_w]$$

For weak bases and their salts:

$$BH^+ \overset{K_a}{\rightleftarrows} B + H^+; \quad K_a = [H^+][B]/[BH^+]$$

In the presence of octanol:

$$B_w \overset{P}{\rightleftarrows} B_o; \quad P = [B_o]/[B_w]$$

Two general conditions govern all titrations:

Material balance (MB)—the total amount of titrated acid or base or of the corresponding salt is equal to the sum of all the components into which it separates.

Ionic balance (IB)—the total concentration of all positively charged ions is equal to the total concentrations of all negatively charged ions.

With these relationships and conditions in mind, the derivation of an equation for P and P' is provided below for each of the four cases. In the derivations below

$$P = \frac{B_o V_w}{V_o B_w}, \therefore B_o + B_w = B_w \left(\frac{V_o P}{V_w} + 1 \right) = f B_w$$

$$P = \frac{HA_o V_w}{HA_w V_o}, \therefore HA_o + HA_w = HA_w \left(\frac{V_o P}{V_w} + 1 \right) = f HA_w$$

Case 1

$$\begin{aligned} MB: \quad V_e &= B_o + B_w + BH^+ \\ [V_e] &= f[B] + [BH^+] \\ \therefore [B] &= (1/f)([V_e] - [BH^+]) \end{aligned}$$

$$\begin{aligned} IB: \quad [H^+] + [BH^+] &= [Cl^-] + [OH^-] \\ \therefore [BH^+] &= [X] - [H^+] + [OH^-] \end{aligned}$$

Whence

$$K_a = \frac{[H^+]}{f} \left(\frac{[V_e] - [X] + [H^+] - [OH^-]}{[X] - [H^+] + [OH^-]} \right) = \frac{K'}{f} \tag{3-1}$$

$$P = (V_w/V_o)(10^{pK_a - pK'} - 1) \tag{3-1a}$$

Also, since

$$P' = \frac{[B_o]}{[B_w] + [BH^+]} \quad \text{and} \quad [B_o] = P[B_w]$$

it is easily shown that

$$P' = \frac{P}{1 + 10^{pK_a - pH}} \tag{3-1b}$$

Case 2

$$\begin{aligned} MB: \quad V_e &= HA_o + HA_w + A^- \\ [V_e] &= f[HA] + [A^-] \\ [HA] &= (1/f)([V_e] - [A^-]) \end{aligned}$$

IB: $[H^+] + [Na^+] = [Cl^-] + [A^-] + [OH^-]$
$$[A^-] = [V_e] - [X] + [H^+] - [OH^-]$$

Whence

$$K_a = f[H^+]\left(\frac{[V_e] - [X] + [H^+] - [OH^-]}{[X] - [H^+] + [OH^-]}\right) = fK' \qquad (3\text{-}2)$$
$$P = (V_w/V_0)(10^{pK'-pK_a} - 1) \qquad (3\text{-}2a)$$
$$P' = P/(1 + 10^{pH-pK_a}) \qquad (3\text{-}2b)$$

Case 3

$$MB: \quad V_e = HA_o + HA_w + A^-$$
$$[V_e] = f[HA] + [A^-]$$
$$[HA] = (1/f)([V_e] - [A^-])$$
IB: $[H^+] + [Na^+] = [A^-] + [OH^-]$
$$[A^-] = [X] + [H^+] - [OH^-]$$

Whence

$$K_a = f[H^+]\left(\frac{[X] + [H^+] - [OH^-]}{[V_e] - [X] - [H^+] + [OH^-]}\right) = fK' \qquad (3\text{-}3)$$
$$P = (V_w/V_0)(10^{pK'-pK_a} - 1) \qquad (3\text{-}2a)$$
$$P' = P/(1 + 10^{pH-pK_a}) \qquad (3\text{-}2b)$$

Case 4

$$MB: \quad V_e = B_o + B_w + BH^+$$
$$[V_e] = f[B] + [BH^+]$$
$$[B] = (1/f)([V_e] - [BH^+])$$
IB: $[H^+] + [NA^+] + [BH^+] = [Cl^-] + [OH^-]$
$$[BH^+] = [V_e] - [X] - [H^+] + [OH^-]$$

Whence

$$K_a = \frac{[H^+]}{f}\left(\frac{[X] + [H^+] - [OH^-]}{[V_e] - [X] - [H^+] + [OH^-]}\right) = \frac{K'}{f} \qquad (3\text{-}4)$$
$$P = (V_w/V_0)(10^{pK_a-pK'} - 1) \qquad (3\text{-}1a)$$
$$P' = P/(1 + 10^{pK_a-pH}) \qquad (3\text{-}1b)$$

Equation (3-1) is easily solved for $[H^+]$ by recalling that $[OH^-] = K_w/[H^+]$.

Cross multiplication gives (brackets and charges omitted)

$$fK_aX - HfK_a + fK_a\frac{K_w}{H} = HV_e - HX + H^2 - K_w$$
$$HfK_aX - H^2fK_a + fK_aK_w = H^2V_e - H^2X + H^3 - HK_w$$
$$H^3 + H^2(V_e - X + fK_a) - H(fK_aX + K_w) - fK_aK_w = 0 \qquad (3\text{-}5)$$

Similarly, Eq. (3-2) gives

$$H^3 + H^2[V_e - X + (K_a/f)] - H[(K_a/f)X + K_w] - (K_a/f)K_w = 0 \quad (3\text{-}6)$$

Equation (3-3) gives

$$H^3 + H^2\left(X + \frac{K_a}{f}\right) - H\left(\frac{K_a}{f}(V_e - X) + K_w\right) - \frac{K_a}{f}K_w = 0 \quad (3\text{-}7)$$

and Eq. (3-4) gives

$$H^3 + H^2(X + fK_a) - H[fK_a(V_e - X) + K_w] - fK_aK_w = 0 \quad (3\text{-}8)$$

Equations (3-5)–(3-8) are of the form $f(H^+)_n$:

$$f(H^+)_n = H^3 + H^2B + HC + D = 0$$

The first derivative of this is

$$f'(H^+)_n = 3H^2 + 2BH + C = 0$$

Using the Newton-Raphson method, the solution of these equations is an iterative process whereby each new approximation of the answer is

$$[H^+]_{n+1} = [H^+]_n - (f[H^+]_n/f'[H^+]_n) \quad (3\text{-}9)$$

B. Equivalent Volume (V_e)

The equivalent volume (V_e) is required for each calculation of the pK_a or pK'. V_e can be calculated from the normality of the titrant and the sample weight and equivalent weight of the compound titrated. However, one must assume that the compound is pure and that there are no errors of measurement. This is often not the case. In practice, the preparation of the solution and the measurement of a suitable aliquot are procedures that may introduce errors. A more precise determination of the pK_a usually results when both V_e and pK_a are calculated from the same titration data. Details of such calculations that are involved in the programs of this chapter are described below.

1. V_e by the Kolthoff Method

The calculation of V_e by the Kolthoff method involves measuring the slopes between each successive pair of the last four points on the titration curve (Fig. 3-3) [7]. The end of the titration is arranged experimentally so that the distances (ΔX) between successive volumes of titrant (X) are equal. Thus, the three slopes to be calculated are

$$S_n = (pH_n - pH_{n-1})/\Delta X$$
$$S_{n-1} = (pH_{n-1} - pH_{n-2})/\Delta X$$
$$S_{n-2} = (pH_{n-2} - pH_{n-3})/\Delta X$$

Absolute values of the slopes are used since the calculation is the same whether the pH values are increasing or decreasing during the titration.

$$V_e = \frac{(S_{n-1} - S_{n-2})\Delta X}{2S_{n-1} - S_{n-2} - S_n} + X_{n-2} \qquad (3\text{-}10)$$
$$\Delta X = X_{n-1} - X_{n-2}$$

This method is fast and accurate when the distances ΔX are small and the slope S_{n-1} is large compared to S_n and S_{n-2}. Thus, the method succeeds when there is an easily discernible graphical end point and it fails otherwise.

2. V_e and pK_a by Iteration

For the calculation of the equivalent volume (V_e) by the iteration method [11], program 32 is designed to use an estimated value of V_e (V'_e) for the first calculation of the standard deviation (SD) of the pK_a values. In calculating the pK_a values, each X value is examined to be sure it falls within the range $0 < X < V_e$. Using this value of V_e as a beginning point, and from now on always using the same set of data points, program 2 now adds incremental values of 0.01 ml to V_e. Each pK_a value is calculated for each data point for each successive value of V_e using Eq. (3-2c) or (3-3c) (see Section C.1). It is obvious that with a pocket calculator this requires considerable time—15 to 30 min are required. To aid the experimenter, however, each successive value of V_e and SD is flashed momentarily during the calculation or it may be printed, if desired. Thus, it is possible to watch the calculator "narrow in" on the correct answer, go past, and then back up one unit and print out the result. Each SD is calculated using N-1 degrees of freedom using the relation

$$SD = \left\{ \frac{\Sigma pK^2 - [(\Sigma pK)^2]/n}{n - 1} \right\}^{1/2} \qquad (3\text{-}11)$$

where n is the number of data points.

3. Confidence Interval

The search method just described for finding V_e and pK_a simultaneously from the same set of data points depends on making a series of successive one-dimensional searches on a set of orthogonal coordinates. As explained very recently by Gordon [15], a "one-dimension" search means that we choose one value for V_e and proceed along its coordinate until we find a value of pK_a that provides a minimum SD. Then we repeat the process along a new V_e coordinate and so on, until an overall minimum SD is located, thus providing final estimates for both V_e and pK_a. As explained by Gordon [15] and recognized earlier by Barry and Meites [10] the statistical measures of confidence intervals (CI) for V_e and pK_a determined in this way involve nonlinear regression analysis. For such an analysis Eqs. (3-2c) and (3-3c) are cumbersome, especially when solved with a calculator program. However, estimates based on a simplified equation

$$pH = pK + \log[(V_e - X)/X] \qquad (3\text{-}12)$$

in which the dissociation of water is not considered, applies as a first approximation. Using this equation, appropriate expressions (3-13) and (3-14) have been derived for the calculation of the 95% confidence limits of V_e and pK_a.

$$CI(\hat{V}_e) = \hat{V}_e \pm t \left[\frac{\sigma^2}{\Sigma(w_i - \overline{w}_i)^2} \right]^{1/2} \qquad (3\text{-}13)$$

$$CI(\overline{p}\hat{K}_a) = \overline{p}\hat{K}_a \pm t \left[\frac{\sigma^2}{\Sigma(w_i - \overline{w}_i)^2} \right]^{1/2} \cdot \left(\frac{\Sigma w_i^2}{N} \right)^{1/2} \qquad (3\text{-}14)$$

Here

$$w_i = \frac{1}{(V_e - X_i) \ln 10}$$

$$\sigma^2 = \frac{\Sigma(pK)^2 - n(\overline{p}K)^2}{n - 2}$$

Normally, Student's t values must be looked up in a table. However, a simple calculator program has been devised to generate the appropriate value (for $\nu = n - 2$ degrees of freedom) whenever it is required. The t value is calculated as follows (see Chapter 4):

$$t = a' + bx + cx^2 + dx^3 \qquad (3\text{-}15)$$

where

$$x = [1 - (1/\nu)]^{-1}, \qquad c = 1.0056,$$
$$a = 0.0086, \qquad\qquad d = -0.269$$
$$b = 1.2134,$$

The error of t thus obtained is ± 0.001 for values of $0 < \nu < 20$.

4. Limitations of the Iteration Method

Worked examples below will illustrate the advantages and limitations of the calculator programs. When the end point is sharp and increments between successive volumes of titrant are small (0.2 ml or less), the Kolthoff method provides a fast and accurate estimate of the equivalent weight and the pK_a is within 0.05 units of the correct value. If the iteration method is used in such cases, an accurate result is obtained within the limits of the experimental techniques used to weigh the sample, measure the volumes, and determine the pH values.

When the end point is not sharp, as indicated by the absence of a defined peak in the graph of the data entered using program 1, the iteration method usually provides an equivalent volume. Even this method will fail however, if the data used are not accurate, or the compound is contaminated with a second titratable component. The iteration method appears to work best when titrant volume increments are equal and symmetrically distributed about $\frac{1}{2}V_e$. The iteration method depends on the detection of a difference in standard deviations of pK_a values calculated with various values of V_e. The SD does not change appreciably if data points are all taken at the center of the titration curve. As shown in the worked examples, a satisfactory determination of the pK_a of p-chloroaniline ($pK_a = 3.98$) has been made at a concentration of 0.01 M.

C. Applications

1. Calculation of pK_a

Equations (3-1)–(3-4) reduce to only two equations, (3-2) and (3-3), when $f = 1$. In this case, Eq. (3-2) provides the pK_a for titration with a strong acid and the pK' for titration with a strong acid in the presence of octanol. Similarly, Eq. (3-3) provides the pK_a for titration with a strong base and the pK' for titration with a strong base in the presence of octanol.

For titration with strong acid, Eq. (3-2) becomes

$$pK_a = pH + \log \frac{[X] - [H^+] + [OH^-]}{[V_e] - ([X] - [H^+] + [OH^-])} \qquad (3\text{-}2c)$$

For titration with strong base, Eq. (3-3) becomes

$$pK_a = pH - \log \frac{[X] + [H^+] - [OH^-]}{[V_e] - ([X] + [H^+] - [OH^-])} \qquad (3\text{-}3c)$$

For the calculator programs

$$[X] = \frac{XN}{V_i + X} \; ; [V_e] = \frac{V_e N}{V_i + X}$$

$$[H^+] = \text{inv } \log(-pH); [OH^-] = \text{inv } \log(pH - 14)$$

Here all volumes are expressed in milliliters; $K_w = 14.00$ at 25°C. Other values of K_w should be used for other temperatures (14.17 at 20°C).

2. Equivalent Weight

The equivalent weight of a compound is easily calculated from the V_e obtained by titration

$$W_e = \frac{W_t}{V_e N} \tag{3-16}$$

wherein W_t is the weighed amount of sample titrated (in milligrams), V_e is the equivalent volume of titrant (in milliliters), and N is the normality of the titrant. The confidence interval

$$CI(W_e) = \frac{CI(V_e) \cdot W_e}{V_e} \tag{3-17}$$

3. Degree of Ionization

The degree of ionization α is the ratio of the concentration of ions to the total concentration of titrated compound.

$$\alpha = \frac{[A^-]}{[HA] + [A^-]}$$

It is easily shown [1, p. 100] that

$$\alpha = [10^{pK_a - pH} + 1]^{-1} \quad \text{(protonated ion)} \tag{3-18a}$$
$$\alpha = (10^{pH - pK_a} + 1)^{-1} \quad \text{(nonprotonated ion)} \tag{3-18b}$$

4. Solubility

The solubility of the un-ionized species formed during a titration is given by the fraction that is un-ionized multiplied by the total concentration, i.e., by $(1 - \alpha)C$.
   Thus

$$[S] = \frac{V_e N}{(V_i + X)(10^{pK_a - pH} + 1)} \tag{3-19a}$$

when the ion is protonated and when the ion is not protonated,

$$[S] = \frac{V_e N}{(V_i + X)(10^{pH - pK_a} + 1)} \tag{3-19b}$$

5. *Partition Coefficient*

There are two equations, (3-1a) and (3-2a), for partition coefficient. The calculator programs select Eq. (3-1a) when the ion is protonated and Eq. (3-2a) when the ion involved in the titration is not protonated. It is clear from Eqs. (3-1) to (3-4) that titrations in the presence of and in the absence of octanol are fully accounted for by the Henderson–Hasselback equation.

For amino acids, which form a zwitterion at the isoelectric point, $\frac{1}{2}(pK_1 + pK_2)$, Eqs. (3-1a) and (3-2a) do not provide P, but rather the ratio of the concentration of the un-ionized compound in octanol to the *total* concentration in the aqueous phase of zwitterion ($^-AB^+H$) and un-ionized compound (HAB). In this case,

$$\frac{P}{K_Z + 1} = \frac{V_w}{V_0} (10^{\Delta pK} - 1)$$

wherein $K_Z = [^-AB^+H]/[HAB]$ and ΔpK is $pK_a - pK'$ or $pK' - pK_a$ as appropriate. Equations (3-1b) and (3-2b) for P' enable the corresponding apparent partition coefficients to be calculated at the given pH. This calculation offers an advantage in studies of biological systems where P' may be determined for compounds and related to the passage of the compounds through biological membranes.

6. *Apparent Partition Coefficient*

Equations (3-1b) and (3-2b) are selected by the calculator programs depending on whether the ion is protonated or nonprotonated, respectively.

7. *Titration Curve*

Equations (3-5)–(3-8) are used to calculate the hydrogen ion concentration (expressed as pH). When octanol is used, a selection is made by the calculator program. Equations (3-5) or (3-7) are selected when the ion is protonated and Eq. (3-6) or (3-8) is selected when the ion is not protonated. When the titrant is an acid, Eq. (3-5) or (3-6) is selected. When the titrant is a base, Eq. (3-7) or (3-8) applies. When octanol is not used, the same selections are made but the equations are simplified because the factor f is unity. The program is designed to cover a very wide range of possibilities so the initial guess for $[H^+]$ is 1 in all cases. Succeeding guesses are changed by a factor of 5 until the calculated $[H^+]$ is less than the guess. When this occurs, the result becomes the guess for the next iteration. Iteration stops when two successive answers do not differ by more than 0.001 pH unit.

Program 34 is designed to calculate individual, selected titration points. It may be made automatic so that large sections of a titration curve, or the entire curve is calculated with any desired increment between calculated titration points. Instructions for making the program automatic and for using the altered program are provided.

III. EXAMPLES

A. Titration of *p*-Cresol with 0.1*N* Sodium Hydroxide

In this example the same set of data is used with each of the programs. The data are experimental results supplied by Adrian Albert and E. P. Serjeant in their book "Ionization Constants of Acids and Bases" [5]. Their data are appropriate for illustrating our programs because the detailed calculation of the pK_a values for each point on the titration curve is supplied by the authors. It is of interest to observe how closely our programs reproduce the original calculations, and then to recalculate the results when some of the parameters are altered.

Table 3-1 presents the original data. A weighed amount (0.0541 gm) of *p*-cresol was dissolved in 47.5 ml of water and titrated with 0.1*N* sodium hydroxide. The pH before addition of titrant was 6.92. The pH following each successive titrant addition is given in Table 3-1, together with the pK_a values calculated by the authors. For the calculation of the mean pK_a, they chose to omit the first and last data points. The weight of *p*-cresol used provided the authors with their estimate of 5.00 ml for the equivalent volume of titrant.

The first column of Table 3-2 illustrates the results obtained with program 31 in which the equivalent volume is assumed to be 5.00 ml. The corresponding printout with the HP-41C calculator is found in Section V.D Table 3-12a, (program 31 RPN), top of column 3. Our calculated pK_as are the same (within 0.01 pK_a unit) as those of the original authors. Note that our program presents the last pK_a first and proceeds backward to the first data point. Albert and Serjeant used the maximum deviation from the mean instead of the standard deviation as provided by our program.

It is sometimes of interest to generate a theoretical titration curve based on the Henderson–Hasselbach equation and to compare the results with experimental data. Program 34 is used for this purpose. The printout for the data of Table 3-1 is presented in Table 3-2, column 2. Each pH is within 0.02 of the experimental value of column 1 except for the calcu-

TABLE 3-1

Titration of p-Cresol with 0.1N KOH[a,b]

Titrant volume (ml)	pH	Calculated pK_a
0	6.92	
0.5	9.19	10.14
1.0	9.55	10.16
1.5	9.77	10.15
2.0	9.97	10.16
2.5	10.14	10.16
3.0	10.29	10.14
3.5	10.46	10.13
4.0	10.64	10.11
4.5	10.84	10.08
5.0	11.08	

$pK_a = 10.14$ (± 0.03) (using inner seven values of the set)

[a] Data from A. Albert and E. P. Serjeant, "Dissociation Constants of Acids and Bases." Barnes & Noble, Inc., New York, distributors, 1962.
[b] 0.0541 gm Compound dissolved in 47.5 ml water at 20°C.

lated initial pH which is 6.06 instead of the experimental value of 6.92. In column 2 the calculated pH values are extended beyond the expected equivalent volume of 5.00 ml. It is interesting to learn whether the Kolthoff method of end-point determination would apply in this case where the end point is not distinct. The theoretical pH values of column 2 were used with program 31 and the slope of the curve at each point was plotted with asterisks (column 3). When label A was pressed to complete the calculation, a pK_a of 10.02 was calculated, but it has a large standard deviation of 0.08 and the equivalent volume of 4.50 ml is not an accurate calculation. This is because the changes in slope near the end point are so small that there is only a small hump in the asterisk curve as the end point is reached and passed. The HP-41C calculator printout corresponding to column 3 of Table 3-1 is found in Section V.D Table 3-12a, (program 31 RPN), bottom of column 3.

It is now of interest to learn whether the iterative program will be more useful with the same data, assuming that the equivalent volume is not known. The results are presented in Table 3-3, columns 1 and 2. Using the raw data of Table 3-1, we note a very distinct trend in the pK_a values at the bottom of column 1. In column 2, we find that when the iteration is

TABLE 3-2

Analysis of Titration Data: p-Cresol and 0.1N NaOH (Table 3-1)

V_e Known	Calculated pH values	V_e by Kolthoff method
BASE TITRATION	T?	BASE TITRATION
N		N
0.1	ION	0.1
VI	B	VI
47.5	OR-	47.5
XS	PKA	XS
0.	10.14	0.
VE	N	PHI
5.	0.1	6.06
	VE	
	5.	
0.5 9.19	P	0.5 9.18
	0.	
	VO	
1. 9.55	0.	1. 9.53
	VI	
1.5 9.77	47.5	1.5 9.76
	X	
	0. 6.06	
2. 9.97	X	2. 9.95
	0.5 9.18 PH	
2.5 10.14	X	2.5 10.12
	1. 9.53 PH	
3. 10.29	X	3. 10.28
	1.5 9.76 PH	
3.5 10.46	X	3.5 10.45
	2. 9.95 PH	
4. 10.64	X	4. 10.63
	2.5 10.12 PH	
4.5 10.84	X	4.5 10.83
	3. 10.28 PH	
	X	5. 11.03
CLCD	3.5 10.45 PH	5.5 11.22
10.08	X	
10.12	4. 10.63 PH	CLCD
10.13		9.86
10.14	4.5 10.83 PH	9.96
10.16		10.01
10.16	5. 11.03 PH	10.04
10.15		10.06
10.16	5.5 11.22 PH	10.07
10.15		10.08
		10.09
5.00 VE		4.50 VE
10.14 PKA		10.02 PKA
0.02 SD		0.08 SD

TABLE 3-3

Analysis of Titration Data: p-Cresol and $0.1N$ NaOH (Table 3-1)

Data input	Iteration ($pK_w = 14.17$)	Iteration ($pK_w = 14.00$)
T?	.0248684816	10.157
B	.0233329504	10.151
H	.0218279096	10.151
0.1	0.020354057	10.149
VI	.0189124909	10.164
47.5	.0175047865	10.163
VE'	.0161332107	10.150
5.	.0148009185	10.161
1.	.0135123693	10.151
0.5	.0122737703	.0059966824
9.19	.0110938733	0.005981648
2.	.0099850407	.0062571299
1.	.0089646082	.0062176161
9.55	.0080566052	.0061806108
3.	.0072929384	.0061461817
1.5	.0067124381	.0061143939
9.77	.0063547099	.0060852917
4.	.0062476886	.0060589087
2.	.0063942181	.0060353065
9.97	.0063686861	.0060145397
5.	.0063454817	.0059966334
2.5	0.006324668	0.005981648
10.14	0.006306241	.0059695896
6.	.0062902544	.0059604866
3.	.0062767189	0.005954386
10.29	.0062656763	.0059513055
7.	.0062571479	.0059512425
3.5	.0062511539	.0059542317
10.46	.0062476886	10.160
8.	.0062467912	10.153
4.	.0062484628	10.153
10.64	10.171	10.150
9.	10.176	10.165
4.5	10.176	10.164
10.84	10.172	10.151
10.	10.184	10.162
10.082	10.181	10.152
10.116	10.166	.0059512425
10.132	10.176	
10.138	10.165	5.0050
10.156	.0062467912	0.0217
10.157		10.16
10.146	5.1690	0.01
10.158	0.0311	
10.149	10.17	
	0.01	

complete, there is no real trend in the pK_a values. The equivalent volume is calculated to be 5.17 ml and the pK_a is 10.17 ± 0.01. Program 32 assumed that the temperature of the titration is 20°C. To see the effect of a change in temperature to 25°C where the K_w value is 14.00 instead of 14.17, the iterative calculation was repeated with the altered K_w value

(Table 3-3, column 3). The corresponding printout with the HP-41C calculator is found in Section V.D Table 3-15a, (program 32 RPN), column 3. In this case the calculated equivalent volume is 5.01 ml. The pK_a is very nearly the same as before (10.16 ± 0.01).

In Table 3-4, column 1, the pK_a calculation is repeated using program 31 and the new, calculated equivalent volume of 5.169 ml. The results are the same as for the iterative program. Table 3-4, column 2 illustrates the use of program 33 with the same data, wherein each individual pK_a value is calculated as the raw data are entered.

In summary, this example illustrates how each program can be used to advantage for different purposes with the same data set.

B. Titration of *Tris* with 0.1N Hydrochloric Acid

In this example (Figs. 3-1, 3-2) data from the manual titration of a strong base, *tris*, [amino*tris*hydroxymethylmethane, $H_2NC(CH_2OH)_3$] with 0.1N hydrochloric acid are entered directly into the calculator (Table 3-5) using program 31. In our example there was no excess of acid. Had there been an excess, its volume would have been automatically subtracted with each volume entry and the corrected volume would be printed. The initial pH is recorded, not just for the record, but to provide a starting point for the calculation of the first slope. It is usually high, i.e., the asterisk is to the right. The volumes of titrant are entered first and then the corresponding pH values. As the end point is approached, the asterisks begin to move to the right, i.e., at volumes of 3.5 and 4 ml. Volume size is then decreased and the titration continued. At volumes of 4.5, 4.6, and 4.7 ml, the asterisk moves further and further to the right. Finally, when 4.8 ml of titrant have been added, the slope of the titration curve begins to decrease and we know the end point is passed. Pressing label A now begins the calculation. V_e is calculated by the Kolthoff method (Fig. 3-3) and then used for the calculation of the pK_a values for each point for which the volume of titrant is less than V_e. In this case, V_e was calculated to be 4.64 ml so the first pK_a calculated is for a volume X of 4.60 ml. This value, 7.63, is the poorest value of the set. Most of the data, in the central portion of the curve, indicate that the measurement is precise. Albert and Serjeant [5] report a value of 8.18 for this pK_a so the accuracy, in this instance, is not outstanding. In Fig. 3-1 the data set are plotted manually on graph paper. When V_e is found graphically by the method of parallel lines (Fig. 3-2), the result is 4.60 ml. If this value is used with program 33, the calculated pK_a for the points near the center of the titration curve is 8.05 (Table 3-5, column 3). It is usually found that a few points near the center of a titration curve provide the most accurate result.

TABLE 3-4

Analysis of Titration Data: p-Cresol and 0.1N NaOH (Table 3-1)

V_e Revised to iterative result

Program 31		Program 33	
BASE TITRATION		PKA CALCN	
N		TITRANT-	
0.1			OH-
VI		N	
47.5		0.1	
XS		VW	
0.		47.5	ML
VE		VE	
5.169		5.169	ML
0.5		0.5	
9.19		9.19	
		10.17	PKA
1.			
9.55		1.	
		9.55	
1.5		10.18	PKA
9.77			
		1.5	
2.		9.77	
9.97		10.17	PKA
2.5		2.	
10.14		9.97	
		10.18	PKA
3.			
10.29		2.5	
		10.14	
3.5		10.18	PKA
10.46			
		3.	
4.		10.29	
10.64		10.17	PKA
4.5		3.5	
10.84		10.46	
		10.18	PKA
CLCD			
10.17		4.	
10.18		10.64	
10.18		10.18	PKA
10.17			
10.18		4.5	
10.18		10.84	
10.17		10.17	PKA
10.18			
10.17		10.17	PKA
		0.01	SD
5.17	VE		
10.17	PKA		
0.01	SD		

TABLE 3-5

Titration of Tris with 0.1N HCl

Program 31			Program 33		
ACID TITRATION	4.3		PKA CALCN		
N	6.92		TITRANT-		
0.1	*				H+
VI	4.4		N		
40.	6.75			0.1	
XS	*		VW		
0.	4.5			40.	ML
PHI	6.43		VE		
9.74	*			4.6	ML
	4.6				
0.5	5.58			2.	
8.82	*			8.14	
*	4.7			8.03	PKA
1.	3.72				
8.53	*			2.5	
*	4.8			7.97	
1.5	3.4			8.05	PKA
8.31	* CLCD				
*	7.63			3.	
2.	7.94			7.79	
8.14	8.01			8.06	PKA
*	8.02				
2.5	8.04				
7.97	8.04			8.05	PKA
*	8.04			0.02	SD
3.	8.05				
7.79	8.04				
*	8.02				
3.5	7.99				
7.55	7.97				
*	7.90				
4.					
7.24	4.64	VE			
*	7.98	PKA			
4.1	0.11	SD			
7.16					
*					
4.2					
7.06					

C.   Titration of p-Chloroaniline with 0.1N Hydrochloric Acid

In this example, a solution of p-chloroaniline in water is titrated manually with 0.1N standardized hydrochloric acid. The solution is dilute, [61.5 mg (0.49 mM) in 40 ml] (0.012 M), and there is no apparent end point in the titration curve (Fig. 3-4). Program 32 is used to locate the equivalent volume V_e and to calculate the pK_a.

Table 3-6 contains all the pertinent data used in the calculation. The beginning of the titration is defined because excess acid was not used. The

TABLE 3-6

Titration of 4-Chloroaniline with $0.1N$ HCl

Calculator printout	
T?	3.808
A	3.840
N	3.856
0.1	3.846
VI	3.846
40.	3.822
VE'	3.828
4.8	3.805
1.	3.791
0.5	3.758
4.7	.0302089924
2.	.0294628172
1.	.0288338888
4.38	.0283234855
3.	.0279318381
1.5	.0276579922
4.16	.0274997509
4.	.0274536816
2.	.0275152057
3.99	.0275043637
5.	.0274945486
2.5	.0274857654
3.81	.0274780209
6.	.0274713218
3.	.0274656715
3.66	.0274610773
7.	.0274575442
3.5	0.027455077
3.48	.0274536816
8.	.0274533629
4.	.0274541268
3.29	3.764
9.	3.805
4.3	3.828
3.15	3.826
10.	3.830
4.6	3.810
3.	3.817
11.	3.797
	3.783
	3.751
	.0274533629
	4.8690
	0.0474
	3.80
	0.03

printout shows that the titrant is an acid (A), that the normality of the titrant is 0.1 and that the initial volume of the titrated solution is 40 ml. The weighed amount of base was 0.49 mM, which would have required 4.9 ml of $0.1N$ HCl. An initial guess for V_e' is 4.8. The first data point, 0.5 ml, has a pH of 4.7. Subsequent data points can be read from the printout.

The 11th point is not entered. Calculation begins when label A is pressed. The initial pK_a values are fairly constant although there is a trend toward smaller values at the beginning of the titration. The SDs are printed, each in turn as the iteration proceeds. The first eight values become successively smaller, but the ninth value is larger than the eighth value. The flashing display indicates that V_e is increased by increments of 0.01 ml until a value of 4.88 ml is reached. V_e is then decreased by 0.001 increments until it passes 4.87 and reaches 4.868. At this point, SD is larger so the value of 4.869 is selected as the correct value. The printout contains the set of pK_a values for $V_e = 4.869$ ml. Then follows the actual SD of 0.027. Finally, the V_e is printed, followed by its 95% confidence interval, then the mean pK_a, and its confidence interval. Albert and Serjeant obtained a pK_a of 3.93 for a similar titration [5].

D. Partition Coefficients

1. Ketotifen

Ketotifen is selected as the first example because the determination of its partition coefficient illustrates many of the experimental aspects involved in the titration method. Ketotifen ($C_{19}H_{19}NOS \cdot C_4H_4O_4$) is the generic name for the maleate salt of a base that has a low solubility in water. The chemical name of the base is 4,9-dihydro-4-(1-methyl-4-piperidylidene)-10H-benzo[4,5]cyclohepta[1,2-b]thiophen-10-one.

An aqueous solution is prepared containing 27.64 mg of the maleate salt in 5 ml of water. The solution is passed through a column (5 mm × 8 cm) of ion exchange resin (AG 3X 4A). To the eluate is added 0.04 ml of $0.1N$ HCl and the solution diluted with water to 10.0 ml. An aliquot (3.5 ml) is added to 26.5 ml water and titrated with $0.1N$ NaOH. A similar aliquot is diluted in the same manner and 4.0 ml of octanol added before titration with $0.1N$ NaOH. The automatic titration curves are presented in Fig. 3-5.

Since excess HCl is added before the titration is begun, it is a simple matter to mark off the starting volume at (a) using parallel lines. The end point volume is also easily marked off at (b) using parallel lines on the aqueous–octanol curve. The difference (b − a) is V_e, the equivalent volume. The aqueous titration gives a precipitate before the titration is complete at point c. Nevertheless, the central portion of both titration curves is a smooth line. Three points are selected in each curve for calculation of pK_a and pK'. The results are presented in the calculator printout, Table 3-7. The printout is self explanatory. The prompting notes on the program serve as labels when the calculation is complete. Log P for ketotifen

TABLE 3-7

Determination of Partition Coefficients

Ketotifen			Pyridine		
PARTITION COEFF			PARTITION COEFF		
TITRANT-			TITRANT-		
ION-		OH-	ION-		OH-
		NH+			NH+
N	0.1		N	0.1	
VW	30.011	ML	VW	5.389	ML
VE	0.1305	ML	VE	0.194	ML
	0.049			0.081	
	7.93			4.93	
	8.15	PKA		5.07	PKA
	0.069			0.101	
	8.2			5.1	
	8.15	PKA		5.06	PKA
	0.089			0.121	
	8.47			5.28	
	8.15	PKA		5.06	PKA
	8.15	PKA		5.06	PKA
	0.00	SD		0.01	SD
VD			VD		
	4.	ML		10.	ML
VW	30.011	ML	VW	5.389	ML
VE	0.1305	ML	VE	0.194	ML
	0.049			0.081	
	5.45			4.	
	5.66	PKA		4.09	PKA
	0.069			0.101	
	5.7			4.16	
	5.64	PKA		4.09	PKA
	0.089			0.121	
	5.98			4.325	
	5.64	PKA		4.08	PKA
	5.65	PK'		4.09	PK'
	0.01	SD		0.01	SD
	2395.62	P		4.62	P
	3.38	LGP		0.66	LGP
	360.37	P'		4.60	P'
	2.56	LG'		0.66	LG'

in this experiment is found to be 3.38. The corresponding log P' at pH 7.4 is 2.56.

An additional item of information is available from point c of Fig. 3-5. The pH at this point is 8.85 and the volume of titrant, X, is 0.116 ml. The solubility $[S]$ of the base is easily calculated.

$$[S] = \frac{0.1305 \times 0.1}{(30.011 + 0.116)(10^{8.85-8.14} + 1)} = 7.07 \text{ m}M/\text{ml}$$

2. Pyridine

Martin [1] notes that a measure of one's technique is the partitioning of pyridine. For this example, we select a sample of pyridine hydrobromide (Baker, practical) and prepare a stock solution of 45.26 mg in 4.5 ml of water and 0.5 ml of 0.1N HCl. For the titrations 0.35 ml aliquots are added to 5.0 ml water and 10.0 ml octanol for the titration in the presence of octanol. The curves have a similar appearance to those in Fig. 3-8 except that now there is no precipitate. The calculator printout is presented in Table 3-7. Our value of 0.66 is within the experimental error quoted by Martin (0.65 ± 0.01) for pyridine.

3. Benzoic Acid

For benzoic acid, two values for log P are provided in the literature: 1.87 (25) and 2.03 (26). Our result is 2.00 and 1.98 in a duplicate determination (Table 3.8). The acid (Baker, analyzed) (27.25 mg) is dissolved in 2.5 ml of water and 2.5 ml of 0.1N NaOH. For the titrations aliquots of the stock solution are acidified with 0.1N HCl and the titrations carried out as before with 0.1N sodium hydroxide. For these titrations the water–octanol volume ratios are approximately 10.5 ml/15.0 ml.

4. p-Aminobenzoic Acid

p-Aminobenzoic acid is a special case that illustrates one of the advantages of the titration method. It is usually necessary to use a buffered system to be certain that it is the neutral, undissociated compound that is being partitioned [20]. The measurement presented no difficulties by using the titration method. A solution is prepared of 32.76 mg of the compound in 2.2 ml of 0.1N HCl and 2.8 ml of water. The titrations are carried out as before using approximately 5 ml of water and 10 ml of octanol (see Table 3-8). Our result (log P = 0.88) compares well with the value of 0.83 reported in reference [20]. Our value (log P' = −1.86) may be compared with a reported value of −1.62 for this constant [2].

TABLE 3-8

Further Examples of Partition Coefficient Determinations

	Benzoic acid (1)			Benzoic acid (2)			p-Aminobenzoic acid		
	PARTITION COEFF TITRANT-			PARTITION COEFF TITRANT-			PARTITION COEFF TITRANT-		
ION-		OH- OR-			OH- OR-			OH- OR-	
N	0.1			0.1			0.112		
VW	10.674	ML		10.713	ML		5.13	ML	
VE	0.205	ML		0.23	ML		0.113	ML	
	0.086			0.107			0.04		
	3.92			3.95			4.43		
	3.95	PKA		3.92	PKA		4.66	PKA	
	0.106			0.127			0.05		
	4.06			4.07			4.57		
	3.95	PKA		3.91	PKA		4.65	PKA	
	0.126			0.147			0.06		
	4.22			4.21			4.73		
	3.96	PKA		3.91	PKA		4.66	PKA	
	3.95	PKA		3.91	PKA		4.66	PKA	
	0.00	SD		0.01	SD		0.01	SD	
VD	15.	ML		15.	ML		10.	ML	
VW	10.674	ML		10.713	ML		5.13	ML	
VE	0.205	ML		0.23	ML		0.113	ML	
	0.086			0.107			0.04		
	5.93			5.99			5.6		
	6.07	PKA		6.05	PKA		5.86	PKA	
	0.106			0.127			0.05		
	6.12			6.15			5.75		
	6.09	PKA		6.06	PKA		5.85	PKA	
	0.126			0.147			0.06		
	6.29			6.32			5.9		
	6.09	PKA		6.07	PKA		5.85	PKA	
	6.08	PK'.		6.06	PK'		5.85	PK'	
	0.01	SD		0.01	SD		0.01	SD	
	96.04	P		101.02	P		7.56	P	
	1.98	LGP		2.00	LGP		0.88	LGP	
	0.03	P'		0.03	P'		0.01	P'	
	-1.46	LG'		-1.48	LG'		-1.86	LG'	

E. Calculation of pH (Titration Curves)

If Fig. 3-5 and the results presented in Tables 3-7 and 3-8 are accurate, it should be possible to use the resulting physical constants and prepare theoretical titration curves for comparison. This was done for ketotifen and the results are presented in Table 3-9 for the curve with octanol first and then for the aqueous curve. The data are plotted on the graph of Fig. 3-5. There is a very slight difference between the curves below 0.03 ml titrant but they differ most after the end point. The aqueous curve is different because it contains a precipitate. The reason for the difference in the octanol curve (after the end point is passed) is not known in this instance. However, such discrepancies in similar curves have been found from the presence of carbon dioxide in the water. The close similarity of the theoretical and actual curves lends strong support to the validity of this useful technique.

IV. INSTRUCTIONS

A. pK_a—Programs 311, 312

Partition 479.59. (For RPN program see also notes p. 119.)

Step 1 *Insert programs* 311, 312 (Tables 3-11 and 3-12 or 3-11a and 3-12a). The program has two versions, one for an acid titrant and one for a base titrant. Be sure to use the appropriate version.

Step 2 *Initialize.* If V_e is known, press D'. If V_e is to be calculated by the program, press E'. The caption "Base Titration" or "Acid Titration" will be printed. The letter N is printed asking for the normality of the titrant.

Step 3 *Enter constants.* Enter the normality (N) of the titrant and press R/S. The number will be printed and the letters V_i will appear, requesting the initial volume. Enter the volume (V_i) and press R/S. The number is printed and the letters X_s appear requesting the volume of excess titrant. When entered, this volume will be automatically subtracted from all subsequent volume entries. If there is no excess, enter 0 and press R/S.

If D' is pressed to initialize, the equivalent volume V_e is requested next. The number is entered, R/S is pressed, the number is printed, and the paper advances (see Table 3-2, column 1).

If E' is pressed to initialize, the initial pH is requested by the letters PH_i. This is entered, R/S is pressed, the number is printed, and the paper advances (Table 3-2, column 3).

Step 4 *Enter raw data.* The volume of titrant is entered, R/S is pressed. The number is printed.

TABLE 3-9

Generation of Titration Curves[a]

Base (pK_a = 8.15, P = 2395.62) titrated with 0.1 N HCl

With octanol			Without octanol		
T?			VD		
	B			0.	
ION			VI		
	NH+			30. 011	
PKA			X		
.	8. 15			0.	
N				5. 76	PH
	0. 1		X		
VE				0. 01	
	0. 1305			7. 07	PH
P			X		
	2395. 62			0. 02	
VD				7. 41	PH
	4.		VI		
VI			X		
	30. 011			0. 05	
X				7. 94	PH
	0.		X		
	4. 52	PH		0. 09	
X				8. 48	PH
	0. 01		X		
	4. 76	PH		0. 1	
X				8. 64	PH
	0. 02		X		
	4. 98	PH		0. 11	
X				8. 83	PH
	0. 05		X		
	5. 45	PH		0. 12	
X				9. 07	PH
	0. 09		X		
	6. 00	PH		0. 13	
				9. 37	PH
X			X		
	0. 12			0. 14	
	6. 71	PH		9. 65	PH
X					
	0. 13				
	7. 89	PH			
X					
	0. 14				
	9. 50	PH			

[a] See Figure 3-5.

The pH is entered and R/S is pressed. If D' is used, the number is printed and the paper advances, ready for the next data point. If E' is used, the number is printed as before but this is followed by an asterisk. The position of the asterisk to the right, or to the left, informs the operator that the curve is steep or of low slope, respectively.

Data points are entered in succession until all points are entered. As many as 19 data points may be entered and stored. If E' is used and there is a pronounced peak in the slope curve (see Table 3-5, column 2), it is only necessary to enter one point past the maximum.

Step 5   *Calculate*. Press A. The heading CLCD is printed, followed by the individual pK_a values (beginning with the last) for each data point whose volume is less than V_e. There is a space and then the equivalent volume (V_e), the pK_a, and its standard deviation (SD) are printed and labeled.

Step 6   *Excess titrant*. Steps 1 to 5 with E' can be used to find the excess titrant if there is a clear break at the beginning of the curve. The value of V_e is printed but it can be saved by the sequence: RCl 21, $X \leftrightarrows t$. E' (or D') is now pressed to begin again and when X_s appears it is recalled with $X \leftrightarrows t$.

B. pK_a and V_e (Iteration)—Programs 321, 322

Partition 559.49 (5 Op 17). (For RPN program see also notes p. 133.)

Step 1.   *Insert programs* 321, 322, 323. (Tables 3-14, 3-15 and 3-16 or 3-14a and 3-15a)

Step 2.   *Initialize*. Press E'. The printer asks for the nature of the titrant (T?).

Step 3.   *Enter conditions*. Press A' for acid or B' for base. A (or B) is printed.

Step 4.   *Enter constants*. The letter N is printed after A or B. Enter the normality of the titrant and press R/S. The normality is printed and V_i appears. Enter the initial volume of the titrated solution and press R/S. The initial volume is printed and the printer asks for the first guess of the equivalent volume V'_e. Enter the guess and press R/S. The V'_e is printed and the number 1 appears to ask for the first data point. *Enter raw data*.

Step 5.   *Calculate*. Press A. The calculation begins by calculating a pK_a value for each data point using V'_e. These are printed beginning with the last point. This is followed by the standard deviation (SD) for this data set. V_e is then increased by 0.01 ml and the calculation is repeated. This time the new V_e and new pK_a are briefly flashed and the new SD is printed. Unless the new SD is larger than the previous SD, the iteration continues in this manner until the new SD is larger than the previous SD. Now the iteration continues but each new V_e is smaller than the previous V_e by 0.001 ml. When the new SD is again larger than the previous SD, the iteration stops. The previous V_e is selected as the result, pK_a values are printed again for each point (with a volume less than V_e), and then the SD for this data set is printed. The paper advances and the summary results are printed: V_e, 95% confidence limit, mean pK_a, and its 95% confidence limit (see Table 3-3). Note that the 95% confidence limits are larger than the SDs and depend on the number of data points used in the calculation.

C. Partition Coefficient—Programs 331, 332, 333

Partition 799.19 (2 Op 17)

Step 1. Insert programs 331, 332, and 333. (Tables 3-18, 3-19 and 3-20).

Step 2. *Initialize.* There are four choices for initialization.

(a) The complete calculation: Calculate pK_a and pK' and then the partition coefficients. Press E'.

(b) pK_a is known. pK' and the partition coefficients are calculated. Press E.

(c) Both pK_a and pK' are known. The partition coefficients are to be calculated. Press C.

(d) Only pK_a (or pK') is to be calculated. Press D. It is sometimes an advantage to calculate pK_a and pK' separately and then the coefficients. This is especially true when there is a precipitate and only a part of the curve is to be used—it is often difficult to know how close to the beginning or how close to the point of precipitation to go until the constancy of the pK_a values is actually observed.

Step 3. *Enter conditions and constants* as prompted.

Titrant: Press A' for acid or B' for base.

Ion: Press C' for nonprotonated ion (OR^-) or D' for protonated ion (NH^+).

N, normality of titrant, R/S after entry; V_w, volume of water (initial), R/S after entry; V_e, equivalent volume, R/S after entry; V_o, volume of octanol, R/S after entry.

Step 4. *Enter raw data* when paper advances. The volume of titrant and pH of the solution are entered for each data point. R/S is pressed after each entry. As each pH is entered and R/S is pressed, the calculation proceeds automatically and pK_a is printed and labeled.

Step 5. *Calculate* when the desired number of pK_a values have been accumulated, label A is pressed to calculate the mean pK_a and its SD. If pK_a has already been calculated, this operation produces pK' (labeled as such) (see Table 3-7). Calculation continues automatically and P, $\log P$, P' and $\log P'$ are printed and labeled. If operation 2c (label C) has been used, calculation proceeds automatically when the V_e is entered and R/S is pressed.

D. pH Calculation—Programs 341, 342

Partition 479.59

Step 1. *Insert programs* 341, 342. (Table 3-22, and 3-23 or 3-22a)

Step 2. *Enter conditions and constants* as prompted.

T?: Press A′ for acid titrant or B′ for base titrant.

ION: Press C′ for nonprotonated ion (OR^-) or D′ for protonated ion (NH^+).

pK_a: Enter and press R/S.

N: Enter normality of titrant, press R/S.

V_e: Enter equivalent volume, press R/S.

P: Enter partition coefficient (if none is involved, enter 0), press R/S.

V_0: Enter octanol volume (if none is involved, enter 0), press R/S.

V_i: Enter initial volume of aqueous solution, press R/S.

X: Enter desired titrant volume, press R/S.

Calculation begins automatically and may continue for 2.5 min. When calculation is complete, the pH is printed and labeled.

Step 3. *Repeat calculations.* If only the titrant volume is changed, press label B. The printer prints X and when the new titrant volume is entered, calculation automatically begins. This program is slow because it is designed to handle all situations. Volume 0, -1, or $+10$ can be entered in any order. See Section VIII.A for additional comments.

If the octanol volume and/or the aqueous solution volume is changed, press A. In this case, only V_0, V_i, and X are repeated (see Table 3-9, column 2).

When the program is made automatic as described in Section VIII.B, it is only necessary to enter the titrant volume X of the first titration point. Subsequent points are calculated at the titrant volume increments specified by the program. Calculation must be stopped by the operator by pressing R/S. The initial point may be a negative value of X and calculation may be continued past the equivalent volume to provide for a smooth titration curve.

Progress of each iteration may be observed if "Pause" is placed in the program between steps 145 and 146. Occasionally, the iteration fails to converge. When this happens, the number 5 at step 415 may be replaced with a smaller number such as 2. Convergence should now occur but the iteration will take longer.

E. Program 34 RPN

Step 1.   Press a for acid titrant or b for base titrant.

Step 2.   Press c for nonprotonated ion (OR^-) or d for protonated ion (NH^+). This step may be omitted if octanol is not involved.

Step 3.   Press E. Enter data in order and press R/S after each entry: pK_a; V_i: (initial volume); N (normality of titrant); V_e (equivalent titrant volume); P (partition coefficient); V_o (volume of octanol); X (titrant volume). Calculation automatically begins. The result of each iteration is momentarily displayed. The final result is left in the display when calculation is complete.

Step 4.   For a new X: enter X, press A.

Step 5.   For a new V_o: enter V_o, press D; enter S, press R/S.

V.   PROGRAM 31—pK_a

A.   Design

This program accepts as many as 19 data points, uses them to calculate the equivalent volume V_e, and then calculates a pK_a value for each data point, prints each pK_a in turn, and finally prints and labels the V_e, the mean pK_a and the standard deviation. As the data points are entered, the slopes of the titration curve are plotted so that one can see visually when the end point has been reached and how steep or shallow the titration curve is in the vicinity of the end point. The end point is calculated by the Kolthoff method using Eq. (3-10) (see Section II). The pK_a values are calculated according to Eqs. (3-2c) or (3-3c) depending on whether the titrant is an acid or a base, respectively. The program is designed in five sections as follows:

Initialization   Pressing label E' clears the memories, resets flag 1, sets the pointer, prints the program title, and prepares for entry of the constants. When V_e is known, label D' is pressed to accomplish these tasks and to set flag 1.

Constants entry   The constants required are normality of the titrant (N), initial volume of the titrated solution (V_i), excess titrant added before the actual titration begins [for instance, to neutralize excess base used to bring an acid into solution (X_s)], and initial pH of the titrated solution (pH_i). If the equivalent volume is known (for instance, because a weighed sample of known molecular weight is used and the end point is not expected to be distinct) then the equivalent volume V_e is entered in place of pH_i (see Table 3-2, column 1). The program is designed so that the user does not need to remember the order of addition of these constants, the printer asks for them. When they are entered into the display, R/S is pressed to print each constant, store it, and ask for the next constant. The constants and intermediate results generated during the calculations are stored in the first 19 data storage registers (see Table 3-10).

Data entry   Raw data: Titrant volume and resulting pH of the solution, are entered next. The data are stored in individual data registers,

TABLE 3-10

Program 31, pK_a

Register contents		
0 excess volume	10 pH_i	20 zero
1 ΣpK	11 S_n	21 V_e
2 $\Sigma(pK)^2$	12 S_{n-1}	22 X_1 (ml)
3 n	13 S_{n-2}	23 pH_1
4 used	14 pointer	:
5 used	15 V_i	:
6 used	16 —	:
7 ΔX	17 X_n (ml)	:
8 —	18 N	58 X_{19}
9 —	19 $pH_n \rightarrow pK_n \rightarrow$ SD	59 $pH_{18} \rightarrow$ save
10 —		pointer

R_{22}–R_{58}, as it is entered with the help of a pointer (R_{14}). There is room for 19 data points. By experience, it is found most informative if the first 10 or so points are evenly spaced and cover the first 90% or so of the titration curve. The last nine points are spaced evenly, much more closely together, and bracket the anticipated end point. As each point is entered, the slope between the last pH and the next to the last pH is calculated and plotted as an asterisk (see Table 3-2). As the end point is approached, the asterisks are printed further and further from the left-hand side, indicating that the curve is becoming steeper. Finally, as the end point is passed, the slopes diminish and the asterisks move to the left again. For calculating the equivalent volume, only one asterisk need be printed after the one that is furthest to the right (see Fig. 3-2). When there is doubt about whether the end point has been reached, several points past the end point may be printed. When the calculation is performed, the program will travel back along the curve until it passes the point of highest slope. When the V_e is known, the asterisks are not needed and the program does not print them (see Table 3-2, column 1).

Calculate   Label A is pressed to begin the calculations. If the V_e is known, calculation proceeds directly to the next section. If not, the program uses the Kolthoff equation (3-10) to calculate the equivalent volume. Using the pointer, the program moves back along the series of data points until the volume entry is reached that immediately precedes the location of the steepest slope. This volume is stored in R_{21}. The Kolthoff equation is then used to calculate the fraction of the distance between this point and the next higher point that corresponds to the inflection in the titration curve. This fraction is added to R_{21} to give the calculated equivalent volume V_e. Once the V_e is known, the pK_a values are calculated for each point that is less than the V_e. R_{20} contains zero and is used as a signal to end the calculations. Details of how the pointer R_{14} operates are provided in annotations to the program. The Henderson–Hasselbach equation [Eq. (3-2c) or (3-3c)] is used to calculate each pK_a. The equation takes the volume of titrant into consideration and corrects as well for the dissociation of water. The pK_a values are printed, each in turn as they are calculated beginning with the point closest to V_e and proceeding back to the beginning. The pK_a values are summed in the statistics registers so that the calculation of the standard deviation is facilitated.

Print results The results are printed and labeled: V_e for equivalent volume, pK_a (as the arithmetic mean), and SD as the standard deviation.

This program uses one flag to skip calculation of V_e if it is already known. Two conditional transfers are used. The operation of these is explained in detail in the annotations to program 31 and illustrated in the decision map (Fig. 3-6).

The value of pK_w used is 14.167 (for 20°C). Other values (such as 14.000 for 25°C) may be substituted at step 227 in the program. The results are expressed to two places of decimals. The fix decimal format may be changed at step 277.

The first pK_a to be calculated is often the least accurate because it is so close to V_e. To omit this point and recalculate, it is necessary to locate the register that stores the last point before V_e. This register number is entered in the pointer R_{14}. The sequence PGM 01, SBR CLR RST will clear the statistics registers (R_1–R_6). Now, if SBR GE is pressed, the last point before V_e will be omitted as the calculation is repeated.

B. Register Contents

Table 3-10 provides the register contents for program 31, pK_a. Raw data are stored sequentially in R_{22}–R_{59}, inclusive, with R_{14} being used as a pointer. When data entry is complete, the pointer is saved in R_{59} since the pH of the last data point is not used in calculations. R_4–R_6 are used by the built-in statistics program of the calculator. R_8–R_{10} and R_{16} are not used. R_{20} is left vacant to serve as a stop signal for program operation.

C. Decision Map

Figure 3-6 is a decision map to illustrate the operation of program 31, pK_a. When label A is pressed, calculation begins by first saving the pointer. This makes it possible, after calculation is complete, to locate the register that contains the last item of data entered. There are two decision

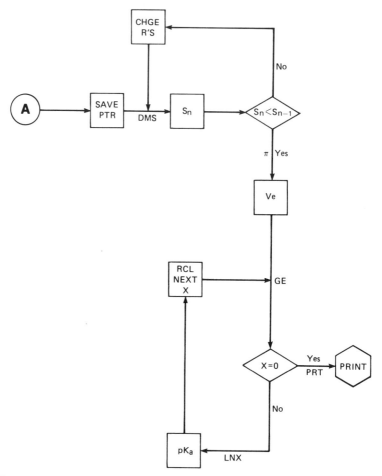

Fig. 3-6 Decision map for program 31, pK_a. When the maximum slope is located (first loop) the equivalent volume is calculated. Then each pK_a is calculated (second loop) and the summary statement is printed.

points. When the calculated slope S_n is less than the slope of the previous point, the equivalent volume V_e is calculated. Then the pK_a is calculated for each data point until the volume X is zero. This occurs when R_{20} is reached (see above).

D. Program Listing

Program 31, pK_a requires a partitioning of 459.59. It may be recorded on two sides of a card. The program listing for each side is provided in

Tables 3-11 and 3-12 for programs 311 and 312, respectively. Label locations are provided in Table 3-12. The program as listed is for titration with base as titrant. The corresponding program, when acid is the titrant, is provided by substituting the keys in the square brackets in the margins of the tables. Printer symbol designations and numbers for the notes that follow the listing are also provided in the margins. The general comments presented in Chapter 1, Section VI apply.

E. Program Notes

1. Programs 311, 312

(1)   For each titration point, the volume of titrant added is entered first. This is always a cumulative value. Excess titrant added before the beginning of the titration curve is stored in R_{00} and is subtracted from the actual burette reading.

(2)   The second data entry is the pH of the solution following the addition of X ml of titrant.

(3)   Flag 1 is set if label D′ is pressed (step 452). Label D′ is pressed if the equivalent volume (V_e) is known and is not to be calculated. In this case, graphical plotting of the slopes is omitted (see step 044).

(4)   The sequence − EXC 10 subtracts the pH of the last data point from the present value and places the present value in R_{10}.

(5)   The pH may be increasing or decreasing, depending on whether the titrant is a base or an acid. The absolute value of the slope is used. It is assumed that equal volumes of titrant are added. It would be relatively easy to add a short sequence to divide by the increment in titrant volume. However, the Kolthoff method of end-point determination is best when the titrant volumes are equal. R_{12} contains the slope between the last and the next to the last data point.

(6)   The factor of 15 is satisfactory for most titrations. A larger number makes the end-point detection more sensitive.

(7)   Op 07 plots an asterisk corresponding to the magnitude of the slope.

(8)   If the slope is too large to fit on the graph, an error condition occurs. Op 19 sets flag 7 if there is an error. If flag 7 is set, operation branches to subroutine IFF (see step 053).

(9)   If flag 7 was not set, the pointer R_{14} is changed and program operation returns to step 001.

(10)   If the slopes are not plotted, the pointer is changed and program operation returns to step 001.

(11)   If an error occurred because the slope was too large to be recorded, the error condition is removed (CLR), and the arbitrary maximum

TABLE 3-11
Program 311 Listing

000	98	ADV	060	86	STF	120	43	RCL	180	14	14
001	91	R/S	061	07	07	121	12	12	181	42	STO
002	99	PRT (1)	062	61	GTO	122	42	STO	182	17	17
003	75	-	063	00	00	123	13	13	183	67	EQ
004	43	RCL	064	34	34	124	43	RCL	184	99	PRT (20)
005	00	00	065	76	LBL	125	11	11	185	01	1
006	95	=	066	11	A (12)	126	42	STO	186	44	SUM
007	72	ST*	067	69	OP	127	12	12	187	14	14
008	14	14	068	00	00	128	61	GTO	188	71	SBR
009	01	1	069	01	1 C	129	88	DMS	189	23	LNX
010	44	SUM	070	05	5	130	76	LBL	190	78	Σ+
011	14	14	071	02	2 L	131	89	⃒ (18)	191	01	1
012	91	R/S	072	07	7	132	01	1	192	22	INV
013	99	PRT (2)	073	01	1 C	133	44	SUM	193	44	SUM (21)
014	72	ST*	074	05	5	134	14	14	194	14	14
015	14	14	075	01	1 K	135	73	RC*	195	61	GTO
016	87	IFF	076	06	6	136	14	14	196	77	GE
017	01	01 (3)	077	69	OP	137	42	STO	197	76	LBL
018	38	SIN	078	01	01	138	21	21	198	23	LNX (22)
019	75	-	079	69	OP	139	02	2	199	43	RCL
020	48	EXC	080	05	05	140	44	SUM	200	15	15
021	10	10 (4)	081	87	IFF	141	14	14	201	85	+
022	95	=	082	01	01	142	73	RC*	202	43	RCL
023	50	I×I (5)	083	77	GE (13)	143	14	14	203	17	17
024	42	STO	084	43	RCL	144	75	-	204	95	=
025	12	12	085	14	14	145	43	RCL	205	35	1/X
026	65	×	086	42	STO	146	21	21	206	65	×
027	01	1 (6)	087	59	59 (14)	147	95	=	207	43	RCL
028	05	5	088	75	-	148	65	×	208	17	17
029	95	=	089	03	3	149	53	(	209	65	×
030	69	OP	090	95	=	150	43	RCL	210	43	RCL
031	07	07 (7)	091	42	STO	151	12	12	211	18	18 (23)
032	69	OP	092	14	14	152	75	-	212	85	+ [-]
033	19	19 (8)	093	73	RC*	153	43	RCL	213	73	RC*
034	87	IFF	094	14	14	154	11	11	214	14	14
035	07	07	095	42	STO	155	54	)	215	42	STO
036	87	IFF	096	19	19 (15)	156	55	÷	216	19	19
037	01	1 (9)	097	76	LBL	157	53	(	217	94	+/-
038	44	SUM	098	88	DMS	158	43	RCL	218	22	INV (23)
039	14	14	099	02	2	159	12	12	219	28	LOG [+]
040	61	GTO	100	22	INV	160	65	×	220	75	-
041	00	00	101	44	SUM	161	02	2	221	53	(
042	01	01	102	14	14	162	75	-	222	73	RC*
043	76	LBL	103	73	RC*	163	43	RCL	223	14	14
044	38	SIN (10)	104	14	14	164	11	11	224	75	-
045	98	ADV	105	75	-	165	75	-	225	01	1 (24)
046	01	1	106	48	EXC	166	43	RCL	226	04	4
047	44	SUM	107	19	19 (16)	167	13	13	227	93	.
048	14	14	108	95	=	168	54	)	228	01	1
049	61	GTO	109	50	I×I	169	95	=	229	06	6
050	00	00	110	42	STO	170	44	SUM	230	07	7
051	01	01	111	11	11	171	21	21	231	54	)
052	76	LBL	112	43	RCL	172	76	LBL	232	22	INV
053	87	IFF (11)	113	12	12	173	77	GE (19)	233	28	LOG
054	20	CLR	114	32	X:T (17)	174	02	2	234	95	=
055	01	1	115	43	RCL	175	22	INV	235	55	÷
056	09	9	116	11	11	176	44	SUM	236	53	(
057	69	OP	117	22	INV	177	14	14	237	94	+/-
058	07	07	118	77	GE	178	29	CP	238	85	+
059	22	INV	119	89	⃒	179	73	RC*	239	43	RCL

114

TABLE 3-12

Program 312 Listing

Addr	Code	Mnem	Note		Addr	Code	Mnem	Note		Addr	Code	Mnem	Note		Addr	Code	Mnem	Note
240	18	18			300	22	INV			360	04	4	I		420	02	2	I
241	65	×			301	79	x̄	(28)		361	69	OP			421	04	4	I
242	43	RCL			302	42	STO			362	03	03			422	69	OP	
243	21	21			303	00	00	(29)		363	03	3	O		423	01	01	
244	55	÷			304	03	3	S		364	02	2			424	69	OP	
245	53	(			305	06	6			365	03	3	N		425	05	05	
246	43	RCL			306	01	1	D		366	01	1			426	91	R/S	
247	15	15			307	06	6			367	00	0			427	99	PRT	
248	85	+			308	69	OP			368	00	0			428	42	STO	
249	43	RCL			309	04	04			369	00	0			429	10	10	
250	17	17			310	43	RCL			370	00	0			430	61	GTO	
251	54	)			311	00	00			371	00	0			431	00	00	
252	54	)			312	58	FIX			372	00	0			432	00	00	
253	95	=			313	02	02			373	69	OP			433	76	LBL	
254	28	LOG	(23)		314	69	OP			374	04	04			434	33	X²	
255	22	INV	[Nop]		315	06	06			375	69	OP			435	04	4	V
256	44	SUM			316	98	ADV			376	05	05			436	02	2	
257	19	19			317	91	R/S			377	69	OP			437	01	1	E
258	43	RCL			318	76	LBL			378	00	00	(31)		438	07	7	
259	19	19	(25)		319	10	E'	(30)		379	03	3	N		439	69	OP	
260	58	FIX			320	22	INV			380	01	1			440	01	01	
261	02	02			321	86	STF			381	69	OP			441	69	OP	
262	99	PRT			322	01	01			382	01	01			442	05	05	
263	92	RTN			323	76	LBL			383	69	OP			443	91	R/S	
264	76	LBL			324	54	)			384	05	05			444	99	PRT	
265	99	PRT	(26)		325	47	CMS			385	91	R/S			445	42	STO	
266	98	ADV			326	29	CP			386	99	PRT			446	21	21	
267	69	OP			327	02	2			387	42	STO			447	61	GTO	
268	00	00			328	02	2			388	18	18			448	00	00	(34)
269	04	4	V		329	42	STO			389	04	4	V		449	00	00	
270	02	2			330	14	14			390	02	2			450	76	LBL	
271	01	1	E		331	22	INV			391	02	2	I		451	19	D'	(35)
272	07	7			332	58	FIX			392	04	4			452	86	STF	
273	69	OP			333	01	1	B [1] A		393	69	OP			453	01	01	
274	04	04			334	04	4	[3]		394	01	01			454	61	GTO	
275	43	RCL			335	01	1	A [1] C		395	69	OP			455	54	)	
276	21	21			336	03	3	[5]		396	05	05			456	00	0	
277	58	FIX			337	69	OP			397	91	R/S			457	00	0	
278	02	02			338	01	01			398	99	PRT			458	00	0	
279	69	OP			339	03	3	S [2] I		399	42	STO			459	00	0	
280	06	06			340	06	6	[4]		400	15	15						
281	22	INV			341	01	1	E [1] D		401	04	4	X					
282	58	FIX			342	07	7	[6]		402	04	4						
283	03	3	P		343	00	0			403	03	3	S					
284	03	3			344	00	0			404	06	6						
285	02	2	K		345	03	3	T		405	69	OP						
286	06	6			346	07	7			406	01	01						
287	01	1	A		347	02	2	I		407	69	OP						
288	03	3			348	04	4			408	05	05						
289	69	OP			349	69	OP			409	91	R/S						
290	04	04			350	02	02			410	99	PRT						
291	79	x̄	(27)		351	03	3	T		411	42	STO						
292	42	STO			352	07	7			412	00	00						
293	19	19			353	03	3	R.		413	87	IFF						
294	58	FIX			354	05	5			414	01	01						
295	02	02			355	01	1	A		415	33	X²						
296	69	OP			356	03	3			416	03	3						
297	06	06			357	03	3	T		417	03	3	P					
298	22	INV			358	07	7			418	02	2	H					
299	58	FIX			359	02	2			419	03	3						

LABELS

Addr	Code	Label
044	38	SIN
053	87	IFF
066	11	A
098	88	DMS
131	89	π
173	77	GE
198	23	LNX
265	99	PRT
319	10	E'
324	54	)
434	33	X²
451	19	D'

TABLE 3-11a

Program 31 RPN Listing

```
opc · ·

01♦LBL "PKA"              54♦LBL 10  (5)         106♦LBL "GE"  (8)
02 "VOL?"                 55 4                    107 2
03 PROMPT                 56 ST+ 14               108 ST- 14
04 RCL 16   (1)           57 RCL IND 14           109 RCL 21
05 -                      58 STO 21               110 RCL IND 14
06 STO IND 14             59 2                    111 X<>Y
07 1                      60 ST- 14               112 X<=Y?
08 ST+ 14                 61 RDN                  113 GTO "GE"
09 "PH?"                  62 RCL IND 14           114 X<>Y
10 PROMPT                 63 X<>Y                 115 STO 17
11 STO IND 14 (2)         64 -                    116 X=0?
12 1                      65 RCL 12               117 GTO "PPT"
13 ST+ 14                 66 RCL 13               118 1
14 GTO "PKA"              67 -                    119 ST+ 14
                          68 *                    120 XEQ "LNX"
15♦LBL F   (3)            69 RCL 12               121 Σ+
16 CLΣ                    70 2                    122 1
17 FS? 01                 71 *                    123 ST- 14
18 GTO "GE"               72 RCL 11               124 GTO "GE"
19 RCL 14                 73 -
20 STO 60                 74 RCL 13               125♦LBL "LNX"  (9)
21 1                      75 -                    126 RCL 15
22 ST- 14                 76 /                    127 RCL 17
                          77 ST+ 21               128 +
23♦LBL 03   (4)           78 SF 04                129 1/X
24 RCL IND 14             79 1                    130 RCL 17
25 2                      80 ST- 14               131 *
26 ST- 14                 81 GTO 03               132 RCL 18
27 RDN                                            133 *
28 RCL IND 14             82♦LBL 11   (6)         134 RCL IND 14
29 -                      83 22                   135 STO 19
30 1                      84 RCL 14               136 CHS
31 ST+ 14                 85 X=Y?                 137 10↑X
32 RDN                    86 GTO 12               138 FS? 00   (10)
33 RCL IND 14             87 1                    139 GTO 04
34 2                      88 ST+ 14               140 +
35 ST- 14                 89 GTO 03               141 GTO 05
36 RDN
37 RCL IND 14             90♦LBL 12   (7)         142♦LBL 04   (10)
38 -                      91 1                     143 -
39 /                      92 ST+ 14
40 ABS                    93 RCL IND 14           144♦LBL 05
41 REGPLOT                94 RCL 10               145 RCL IND 14
42 FS? 04                 95 -                    146 14.167
43 GTO 11                 96 1                    147 X<>Y
44 STO 11                 97 ST- 14               148 -
45 RCL 12                 98 RDN                  149 CHS
46 X>Y?                   99 RCL IND 14           150 10↑X
47 GTO 10                 100 -                   151 FS? 00
48 STO 13                 101 /                   152 GTO 06
49 RDN                    102 ABS                 153 -
50 STO 12                 103 REGPLOT             154 GTO 07
51 1                      104 RCL 60
52 ST+ 14                 105 STO 14              155♦LBL 06   (10)
53 GTO 03                                         156 +
```

TABLE 3-12a
Program 31 RPN Listing and Example

157♦LBL 07	217 AOFF	V_e Known (18)
158 ENTER↑	218 ASTO 16	PROG PKA
159 ENTER↑	219 "ID= "	ID= CRESOL
160 RCL 15	220 ARCL 16	BASE TITRANT
161 RCL 17	221 AVIEW	N= 0.100
162 +	222 "TITRANT"	VI= 47.500
163 1/X	223 ASTO 03	XS= 0.000
164 RCL 21	224 ASHF	VE= 5.000
165 *	225 ASTO 04	10.083
166 RCL 18	226 CLA	10.117
167 *	227 ARCL 03	10.133
168 X<>Y	228 ARCL 04	10.138
169 -	229 PROMPT	10.156
170 /		10.158
171 LOG		10.146
172 FS? 00 (10)	230♦LBL A (14)	. 10.158
173 GTO 08	231 SF 00	10.149
174 ST- 19	232 "ACID "	
175 GTO 09	233 ARCL 03	VE= 5.000
	234 ARCL 04	PKA= 10.139
176♦LBL 08 (10)	235 AVIEW	STD DEV= 0.025
177 ST+ 19	236 GTO 01	
178♦LBL 09 (11)	237♦LBL B (15)	
179 RCL 19	238 "BASE "	PROG PKA
180 VIEW X	239 ARCL 03	ID= CRESOL
181 RTN	240 ARCL 04	BASE TITRANT
	241 AVIEW	
182♦LBL "PRT" (12)	242♦LBL 01 (16)	V_e Unknown (19)
183 ADV	243 "N= "	N= 0.100
184 "VE= "	244 PROMPT	VI= 47.500
185 ARCL 21	245 STO 18	XS= 0.000
186 AVIEW	246 ARCL 18	INIT PH= 6.060
187 MEAN	247 PRA	
188 STO 19	248 "VI= "	
189 "PKA= "	249 PROMPT	
190 ARCL 19	250 STO 15	
191 AVIEW	251 ARCL 15	
192 SDEV	252 PRA	
193 STO 00	253 "XS= "	
194 "STD DEV= "	254 PROMPT	
195 ARCL 00	255 STO 16	
196 AVIEW	256 ARCL 16	
197 ADV	257 PRA	
198 STOP	258 0	9.857
	259 "VE= "	9.959
199♦LBL E (13)	260 PROMPT	10.007
200 CF 00	261 X≠0?	10.041
201 CF 01	262 GTO 02	10.059
202 CF 04	263 "INIT PH= "	10.069
203 ΣREG 03	264 PROMPT	10.080
204 CLRG	265 STO 10	10.088
205 22	266 ARCL 10	
206 STO 14	267 PRA	
207 FIX 3	268 GTO "PKA"	VE= 4.500
208 1		PKA= 10.020
209 STO 01	269♦LBL 02 (17)	STD DEV= 0.078
210 168	270 SF 01	
211 STO 02	271 STO 21	
212 "PROG PKA"	272 ARCL 21	
213 PRA	273 PRA	
214 "ID?"	274 GTO "PKA"	
215 AON	275 END	
216 PROMPT		

slope of 19 is printed with Op 07. The flag 07 is reset and program operation resumes at step 034.

(12)   Label A is pressed to initiate the calculation.

(13)   If flag 1 is set, the V_e is known and program operation omits the calculation of V_e and goes directly to calculate pK_a values at label GE (step 173).

(14)   The pointer is saved in case the data are to be recalled for some other purpose. For instance, a program can be written to print the data in columns in the format of a table.

(15)   The pH of the next to last data point is stored in R_{19}.

(16)   This step is similar to that of note (4) except that now the penultimate slope is calculated.

(17)   The penultimate slope is compared with the slope between the last two points. If it is smaller, it is assumed that the maximum slope is in R_{12}. If it is larger, then the last two slopes are saved in R_{12} and R_{13}, and program operation continues now with calculation of the third slope from the last. In fact, the process continues until there is a sequence of three successive slopes in R_{11}, R_{12}, and R_{13} such that R_{12} is larger than R_{11}. When this is the case, program operation moves to label π at step 131 (see Fig. 3-6).

(18)   Label π calculates V_e by the Kolthoff method using the three slopes in R_{11}, R_{12}, and R_{13} (see Section II). The third volume (X_{n-2} in Fig. 3-3) becomes the temporary V_e and is stored in R_{21}. The Kolthoff formula provides a volume increment to be added to X_{n-2} to obtain the final value of V_e.

(19)   Label GE calculates the pK_a for each point that is less than V_e until all the points have been calculated. Note that R_{20} contains a zero to signal the end of the data storage.

(20)   When all the pK_as are calculated, the program operation transfers to label PRT to print the results (step 265).

(21)   The use of the pointer during the calculations is complex because it is used to locate X and pH.

(22)   Label ln X calculates the pK_a for each point.

(23)   For acids the program is altered as in brackets.

(24)   The value of pK_w is 14.167 at 20°C. For other temperatures other values should be used, for instance, 14.00 at 25°C.

(25)   Each pK_a is printed in turn. Two significant figures are adequate for most experimental work.

(26)   Label PRT prints the final results with appropriate labels.

(27) $\bar{X}$ provides the arithmetic mean of the pK_a values. To be more accurate, each pK_a should be converted to its anti-log before the mean is calculated of the acid-dissociation constants.

(28)   The standard deviation calculated in this way is not strictly correct (see note 27).

(29)   The Op codes interfere with INV $\bar{X}$. It is apparently necessary to store the standard deviation and then recall it after the print register has been filled.

(30)   Label E' initializes the program by resetting flag 1, clearing the memories, clearing the t register (CP), and setting the pointer R_{14}. The program printout is labeled with "Base Titration" or "Acid Titration."

(31)   Constants used in the titration are requested by the printer. When entered into the display, they are printed and stored in appropriate registers.

(32)   It is necessary to enter 0 into the display if there is no excess titrant.

(33)   If flag 1 is set, V_e is known and is entered at label X^2 (step 434). If not, the initial pH is entered at this point to provide data for the first asterisk of the graph.

(34)   RST is not used because it would reset flag 1.

(35)   Label D' sets the flag 1 and then initializes as described in note 30 and subsequent notes.

2.   Program 31 RPN

(1)   Excess titrant volume stored in R_{16} is automatically subtracted from titrant volume entries.

(2)   Entered data are not printed.

(3)   LBL F begins calculation. If flag 1 is set, V_e is known.

(4)   Slopes are calculated and plotted. When the end point is passed, program operation moves to LBL 10 to calculate V_e.

(5) V_e is calculated by the Kolthoff method, Eq. (3-10).

(6)   When all titration points have been examined, program operation moves to LBL 12.

(7)   This sequence plots the first point.

(8)   Statistics are computed.

(9) pK_a values are calculated.

(10)   The appropriate sign is used depending on whether the titrant is an acid or a base.

(11)   The pK_a value is displayed as it is calculated.

(12)   The results are printed.

(13)   LBL E initiates the program for data entry.

(14)   LBL A is pressed if the titrant is an acid.

(15)   LBL B is pressed if the titrant is a base.

(16)   Prompting messages ask for data input.

(17) When V_e is known, flag 1 is set.
(18) Input for this example is provided in Table 3-2, column 1.
(19) Input for this example is provided in Table 3-2, column 3.

VI. PROGRAM 32—pK_a AND EQUIVALENT VOLUME (ITERATION)

A. General Considerations

Program 32 is used when one end of a titration curve is poorly defined. This may occur because a precipitate forms during the titration or because the pK_a of the compound being titrated is less than 4.5 or greater than 9.5. The volumes of titrant must be expressed relative to the end of the titration curve that is defined. When the titration curve is defined only at its end point, the usual calculation is reversed. In this case, the curve is treated as though it was the mirror image with the defined end of the curve at the beginning. Thus, if the titrant is an acid in this case, the program for a base titrant is used, and vice versa. Typical titration curves for which program 32 is most useful are illustrated in Fig. 3-7.

For the calculations, it is assumed that both the pK_a and the equivalent volume V_e are unknown but all other relative parameters involved in the calculation are known. Program 32 explores the two dimensional space of V_e and pK_a values until a value of V_e is found such that the standard deviation SD of the mean pK_a is a minimum. The operator selects an initial trial value of V_e, namely V_e'. Points from the titration curve are then selected, such that titrant volume intervals Δx are equal and distributed symmetrically (as near as is practical) about $\frac{1}{2} V_e$. As many as 13 such data points (x and pH values) are entered into the calculator program. When the points are entered, label A is pressed and calculation begins.

Using V_e' the pK_a is calculated by the program for each entered point that has a volume x less than V_e'. (Initially, of course, all x values are selected to be less than V_e', but as the calculation proceeds, a test is made on each calculation to be sure this is the case. When it is not, that point is excluded from subsequent calculations. The mean pK_a and SD are calculated. A volume increment is then added to V_e' and a new set of pK_a values is calculated. A new SD is found and compared with the previous value. This process is continued with the size and sign of the volume increments being systematically changed. With each iteration, the new SD is compared as before until the minimum SD is located. The V_e that provides the minimum SD is stored and its 95% confidence limits are calculated as well as the mean pK_a and its 95% confidence limits. These results are then printed as a summary statement.

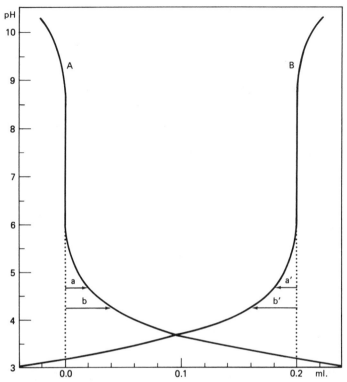

Fig. 3-7 Titration curves with $V_e = 0.2$ ml for (A) The salt of an acid of pK_a 3.60 titrated with $0.1N$ HCl, and (B) the salt of a base of pK_a 3.60 titrated with $0.1N$ NaOH. Curve B is the mirror image of curve A. For curve A, a and b are titrant volumes. For curve B titrant volumes a' and b' are entered into program 32 with acid entered as titrant.

If V_e is very much different from the initial V'_e when the summary statement is printed, a more precise result will be obtained if a new set of data points are selected from the curve, which are more symmetrically distributed about $\frac{1}{2} V_e$ and the iteration is performed again. Naturally, when a precipitate forms early during the titration, it is not possible to select a set of data points symmetrically distributed about $\frac{1}{2} V_e$. In many cases, however, the pK_a calculated from such a partial curve will be quite satisfactory.

B. Design

(1) *Initialization* Pressing E' clears the memories, resets the flags, sets the pointer, enters the Δ's and prepares for entry of the constants by asking (*via* the printer) for the nature of the titrant. A' is then pressed if the

titrant is an acid, or B' if the titrant is a base. A or B is printed accordingly.

The program is written with $\Delta 1 = 0.01$ and $\Delta 2 = 0.001$. These increments may be changed provided the changes occupy the same program space (since direct addressing of subroutines is used).

(2) *Constants*   The constants are called for and printed as they are entered. These are the normality (N) of the titrant and the initial volume (V_i) of the titrated solution (including any excess titrant consumed before the curve begins). If an excess of titrant is to be subtracted from each volume entry, the amount of the excess is entered manually in R_{07}.

(3) *Raw data entry*   The number 1 is now printed to ask for the data of the first titration point. The volume of titrant and the corresponding pH are entered and R/S pressed after each entry, for each point in turn. The numbered points are helpful because one prefers more than six points for acceptable statistics and the program cannot accommodate more than 13 points. It is best to select evenly spaced points evenly distributed about the midpoint of the titration curve ($\frac{1}{2}V_e$). Points too close to the beginning are to be avoided, but some points near the beginning of the curve are required so that SD will change appreciably as V_e is altered by the program search.

(4) *Calculations*   Formulas involved in the calculations are described in Section II. Program operation may be explained by referring to Fig. 3-8. Four flags are used. Flags 1 and 2 govern the sign and magnitude, respectively, of the increment added to V_e with each iteration. Flag 3 causes the pK_a values of the first and last iteration to be printed. Flag 4 signals the summary statement to be calculated and printed.

Calculations begin when label A is pressed. The pointer is set and $R_0–R_6$ are cleared for the statistics calculation. Each point on the titration curve is examined beginning with the last. When the titrant volume X is not less than V_e, the calculation of pK_a is skipped and the next point is selected. During the first iteration, flag 3 is not set so each pK_a is printed as it is calculated. When all points have been examined, $x = 0$ and flag 3 is set. The standard deviation, SD_1 is calculated. Before A is pressed, SD_0 is set equal to 1 by placing 1 in R_{12}. SD_0 is now recalled and since it equals 1, comparison with SD_1 is skipped. The calculated SD_1 is placed in R_{12} to become SD_0 for the next iteration. Flag 2 has not been set so the increment Δ in R_{08} is added to V_e and the second iteration is begun.

After the completion of the second iteration, SD_1 is not unity so it is compared with SD_0 of the first iteration. If SD_1 is less than SD_0, the second iteration has a V'_e, which is closer to the correct result V_e than the first one. Thus, the sign of the Δs is correct. On the other hand, if SD_1 is not less than SD_0, the sign of the Δs in R_{08} and R_{10} is changed. In either

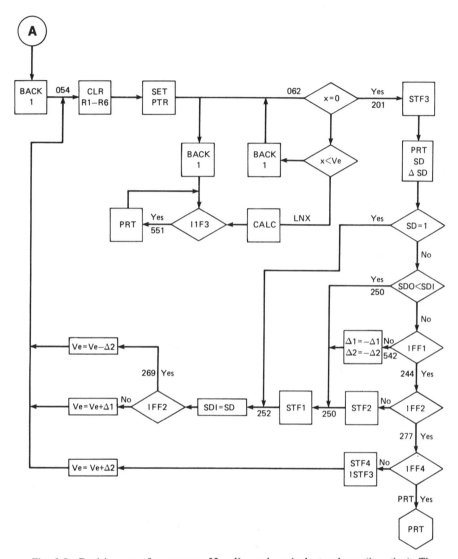

Fig. 3-8 Decision map for program 32, pK_a, and equivalent volume (iteration). The upper set of loops calculates a mean pK_a and its standard deviation (SD). The lower set of loops changes the equivalent volume systematically until a minimum SD is located. Finally the result is printed.

case, flag 1 is set to indicate that the Δs have the proper sign. Iteration continues to add the increment in R_{08} to V'_e until SD_1 is no longer less than SD_0. When this happens, flag 2 is examined to see if it is set. Since it is not set, the program sets flag 2 and now subtracts the smaller Δ in R_{10} from V'_e. Again iteration continues until SD_1 is no longer less than SD_0. Now flags 1 and 2 are both set. Flag 4 is not set so the program sets it, resets flag 3, and then adds a final value of the smaller Δ to V_e. On the last iteration, the program finds flag 3 is set so the pK_a values are printed. Flags 1, 2, and 4 are set so the final results are calculated, and a summary statement is printed.

The calculations for the statistics are explained in Section II. Student's t value is computed by a formula derived in Example III-E using $n - 2$ degrees of freedom.

(5) *Print results*   The summary statement is printed by simply recalling the appropriate registers and printing the contents.

(6) *Labels*   Only one label is used, which is not user-defined. This is PRT for computation and printing of the summary statement (step 293). Of the user-defined labels, E' initializes the program, A' or B' signals that the titrant is an acid or a base, respectively, A begins the calculation, and D' begins calculation again when an error has been corrected.

(7) *Error correction*   Since the iteration can be quite long, 15 to 20 min, the initial set of pK_a values is printed so the operator can determine if there is an obvious error in data entry. If an error is detected, the program can be stopped by pressing R/S. If a volume or a pH is entered in error, the memory in which the erroneous value is located must be found either by trial and error or by printing all the memories by pressing 21 INV LIST. When the correct value is entered, the number 1 is entered into R_{12} and the desired V'_e is entered into R_{21}. Pressing D' now causes the calculation to begin by resetting all the flags and then going to label A.

Progress of the iteration may be followed in a number of ways. The program has PAU instructions at steps 205, 270, and 218. These cause the values of V_e, the mean pK_a, and the difference $SD_1 - SD_0$, respectively, to be displayed momentarily with each iteration, while SD_1 is printed. It is a simple matter to replace any or all of these PAU instructions with PRT or NOP as desired for faster and more efficient program operation. Initially they help the operator to locate program errors. Progress of the iteration may be followed by observing the changes in SD_1 or in the difference $SD_1 - SD_0$. The latter usually becomes smaller as the correct value of V_e is approached. It is important that the initial volume V_i be correct. A curve may be fitted with both V_e and pK_a incorrect if V_i is incorrect.

Nevertheless, this program has consistently given results that are very close to those obtained when the corresponding theoretical curve is calculated with program 34.

C. Register Contents

Program 32 is longer than program 31 and the number of registers available for data points is smaller (13 instead of 19), otherwise the design is similar (see Table 3-13). R_{49} is used to save the pointer, in this program a necessity because the memory contents are used over again with each iteration. R_{20} is vacant as a stop signal; R_4 and R_5 are used, in this program, to store values of ω for the statistics calculations. A calculated Student's t value is stored in R_{00}.

D. Program Listing

Program 32, pK_a and equivalent volume (iteration) requires a partitioning of 559.49 (5 Op 17). It requires two cards (three-card sides) for recording. The program listing is provided in Tables 3-14, 3-15, and 3-16 for programs 321, 322, and 323, respectively. The same program has been designed to handle either acid or base titrant. To enable the program listing to be followed more easily, many of the register designations are provided in the margins of the table as well as the numbers in parentheses for program notes. Label locations are provided in Table 3-16. The general comments of Chapter 1, Section VI, apply.

E. Program Notes

1. Programs 321, 322, 323

(1) Initialization uses RST to reset the flags (step 422). GTO 423 sends program operation back again.

(2) Op 20 places 1 in R_{00}. Recalling R_{00} asks for the raw data of the first data point.

TABLE 3-13

Program 32, pK_a and Equivalent Volume (Iteration)

Register contents		
0 t	10 Δ_2	20 0
1 ΣpK	11 SD_n	21 V_e
2 $\Sigma(pK)^2$	12 SD_{n-1}	22 X_1
3 n	13 V'_e	23 pH_1
4 $\Sigma\omega$	14 pointer	:
5 $\Sigma\omega^2$	15 V_i	:
6 used	16 pointer $\rightarrow \varphi$	46 X_{13}
7 X_s	17 $X_n \rightarrow$ CI (V_e)	47 pH_{13}
8 Δ	18 N	48 —
9 $+1$ or -1	19 $pH_n \rightarrow$ CI (pK_a)	49 save pointer

TABLE 3-14

Program 321 Listing

000	61	GTD		060	42	STD		120	18	18	N	180	17	17	
001	04	04	(1)	061	16	16		121	85	+		181	95	=	
002	23	23		062	00	0		122	43	RCL		182	65	×	
003	69	DP		063	32	X:T		123	09	09		183	01	1	
004	20	20	(2)	064	73	RC*		124	65	×		184	00	0	
005	43	RCL		065	16	16	(8)	125	73	RC*		185	23	LNX	
006	00	00		066	67	EQ		126	16	16	pH	186	95	=	
007	99	PRT		067	02	02		127	42	STD		187	35	1/X	
008	91	R/S		068	01	01		128	19	19		188	44	SUM	
009	75	-		069	43	RCL		129	94	+/-		189	04	04	
010	43	RCL		070	21	21		130	22	INV		190	33	X²	
011	07	07	(3)	071	32	X:T		131	28	LDG		191	44	SUM	
012	95	=		072	73	RC*	(9)	132	75	-		192	05	05	
013	72	ST*		073	16	16		133	43	RCL		193	43	RCL	
014	14	14		074	22	INV		134	09	09		194	19	19	pK_a
015	99	PRT		075	77	GE		135	65	×		195	22	INV	
016	01	1		076	00	00		136	53	(		196	87	IFF	
017	44	SUM		077	85	85		137	73	RC*		197	03	03	
018	14	14		078	02	2		138	16	16		198	05	05	
019	91	R/S		079	22	INV		139	75	-		199	51	51	
020	72	ST*		080	44	SUM		140	01	1		200	92	RTN	
021	14	14	(4)	081	16	16		141	04	4		201	86	STF	
022	99	PRT		082	61	GTD		142	93	NOP	(14)	202	03	03	(16)
023	01	1		083	00	00		143	01	NOP		203	43	RCL	V_e'
024	44	SUM		084	62	62		144	07	NOP		204	21	21	
025	14	14		085	42	STD		145	54	)		205	66	PAU	(17)
026	43	RCL		086	17	17	(10)	146	22	INV		206	79	x̄	pK_a
027	14	14		087	01	1		147	28	LDG		207	66	PAU	
028	42	STD		088	44	SUM		148	95	=		208	22	INV	
029	49	49		089	16	16		149	55	÷		209	79	x̄	SD1
030	61	GTD		090	71	SBR		150	53	(		210	99	PRT	
031	00	00	(5)	091	23	LNX	(11)	151	94	+/-		211	42	STD	
032	03	03		092	44	SUM		152	85	+		212	11	11	
033	68	NDP		093	01	01		153	43	RCL		213	94	+/-	
034	44	SUM		094	33	X²		154	18	18	N	214	85	+	
035	08	08		095	44	SUM	(12)	155	65	×		215	43	RCL	
036	61	GTD		096	02	02		156	43	RCL	V_e'	216	12	12	SDO
037	00	00		097	69	DP		157	21	21		217	95	=	
038	54	54		098	23	23		158	55	÷		218	66	PAU	ΔSD
039	22	INV		099	03	3		159	53	(		219	43	RCL	
040	44	SUM		100	22	INV	(13)	160	43	RCL	V_i	220	12	12	(18)
041	08	08		101	44	SUM		161	15	15		221	32	X:T	
042	61	GTD		102	16	16		162	85	+		222	01	1	
043	00	00		103	61	GTD		163	43	RCL	X	223	67	EQ	
044	54	54		104	00	00		164	17	17		224	02	02	
045	76	LBL		105	62	62		165	54	)		225	52	52	
046	11	A	(6)	106	76	LBL		166	54	)		226	43	RCL	
047	43	RCL		107	23	LNX	(11)	167	95	=		227	12	12	(19)
048	49	49		108	43	RCL		168	28	LDG		228	32	X:T	
049	75	-		109	15	15	V_i	169	65	×		229	43	RCL	
050	02	2		110	85	+		170	43	RCL		230	11	11	
051	95	=		111	43	RCL		171	09	09		231	22	INV	
052	42	STD		112	17	17	x	172	95	=		232	77	GE	
053	14	14		113	95	=		173	22	INV		233	02	02	
054	36	PGM	(7)	114	35	1/X		174	44	SUM		234	50	50	
055	01	01		115	65	×		175	19	19	pK_a	235	87	IFF	
056	71	SBR		116	43	RCL		176	43	RCL	(15)	236	01	01	(20)
057	25	CLR		117	17	17		177	21	21		237	02	02	
058	43	RCL		118	65	×		178	75	-		238	44	44	
059	14	14		119	43	RCL		179	43	RCL		239	68	NDP	

TABLE 3-15

Program 322 Listing

#	Code	Mn		#	Code	Mn		#	Code	Mn		#	Code	Mn	
240	68	NOP		300	16	16	φ	360	55	÷		420	76	LBL	(25)
241	61	GTO		301	01	1		361	43	RCL		421	10	E'	
242	05	05		302	75	-		362	16	16		422	81	RST	(1)
243	42	42		303	43	RCL		363	55	÷		423	22	INV	
244	87	IFF		304	16	16		364	53	(		424	58	FIX	
245	02	02		305	35	1/X		365	43	RCL		425	69	OP	
246	02	02		306	95	=		366	05	05		426	00	00	
247	77	77		307	35	1/X		367	75	-		427	47	CMS	
248	86	STF		308	42	STO		368	43	RCL		428	93	.	
249	02	02		309	00	00		369	03	03		429	00	0	
250	86	STF		310	65	×		370	65	×		430	01	1	
251	01	01		311	01	1		371	79	x̄		431	42	STO	
252	68	NOP		312	93	.		372	32	X:T		432	08	08	Δ1
253	68	NOP		313	02	2		373	33	X²		433	93	.	
254	43	RCL		314	01	1		374	54	)		434	00	0	
255	11	11		315	03	3		375	54	)		435	00	0	
256	42	STO		316	04	4		376	34	ΓX		436	01	1	
257	12	12		317	85	+		377	95	=		437	42	STO	Δ2
258	87	IFF		318	93	.		378	42	STO		438	10	10	
259	02	02		319	00	0		379	17	17	CI	439	02	2	
260	02	02		320	00	0		380	65	×		440	02	2	(26)
261	69	69		321	08	8		381	53	(		441	42	STO	
262	43	RCL		322	06	6		382	43	RCL		442	14	14	
263	08	08		323	85	+		383	05	05		443	01	1	
264	44	SUM		324	43	RCL		384	55	÷		444	42	STO	
265	21	21		325	00	00		385	43	RCL		445	12	12	(27)
266	61	GTO		326	33	X²		386	03	03		446	03	3	T
267	00	00		327	65	×		387	54	)		447	07	7	
268	54	54		328	01	1		388	34	ΓX		448	07	7	?
269	43	RCL		329	93	.		389	95	=		449	01	1	
270	10	10		330	00	0		390	42	STO		450	69	OP	
271	22	INV		331	00	0		391	19	19	CI	451	01	01	
272	44	SUM		332	05	5		392	98	ADV		452	69	OP	(28)
273	21	21		333	06	6		393	43	RCL		453	05	05	
274	61	GTO		334	85	+		394	21	21	Ve	454	91	R/S	
275	00	00		335	43	RCL		395	58	FIX		455	03	3	N
276	54	54		336	00	00		396	04	04	(23)	456	01	1	
277	87	IFF		337	45	Y×		397	99	PRT		457	69	OP	
278	04	04		338	03	3		398	43	RCL		458	01	01	
279	99	PRT		339	65	×		399	17	17	CI	459	69	OP	(29)
280	86	STF		340	93	.		400	99	PRT		460	05	05	
281	04	04		341	02	2		401	03	3	P	461	91	R/S	
282	22	INV		342	06	6		402	03	3		462	42	STO	
283	86	STF		343	09	9		403	02	2	K	463	18	18	
284	03	03		344	94	+/-		404	06	6		464	99	PRT	
285	43	RCL		345	95	=		405	01	1	A	465	04	4	V
286	10	10		346	42	STO		406	03	3		466	02	2	
287	44	SUM		347	00	00	l	407	69	OP	·	467	02	2	I
288	21	21		348	65	×		408	01	01		468	04	4	
289	61	GTO		349	53	(		409	79	x̄	pKa	469	69	OP	
290	00	00		350	53	(		410	58	FIX		470	01	01	
291	54	54		351	43	RCL		411	02	02	(24)	471	69	OP	·
292	76	LBL		352	02	02		412	99	PRT		472	05	05	(30)
293	99	PRT	(21)	353	75	-		413	43	RCL		473	91	R/S	
294	43	RCL		354	43	RCL		414	19	19	CI	474	42	STO	
295	03	03		355	03	03		415	99	PRT		475	15	15	
296	75	-		356	65	×		416	98	ADV		476	99	PRT	
297	02	2	(22)	357	79	x̄		417	98	ADV		477	04	4	V
298	95	=		358	33	X²		418	98	ADV		478	02	2	
299	42	STO		359	54	)		419	91	R/S		479	01	1	E

TABLE 3-16

Program 323 Listing

480	07	7		540	61	GTD	
481	06	6		541	11	A	(35)
482	05	5		542	01	1	
483	69	DP		543	94	+/-	
484	01	01		544	49	PRD	
485	69	DP		545	08	08	
486	05	05	(31)	546	49	PRD	
487	91	R/S		547	10	10	
488	42	STD		548	61	GTD	
489	21	21		549	02	02	
490	99	PRT		550	50	50	
491	61	GTD		551	58	FIX	
492	00	00		552	03	03	
493	03	03		553	99	PRT	
494	76	LBL		554	22	INV	
495	16	A'	(32)	555	58	FIX	
496	01	1		556	92	RTN	
497	94	+/-		557	00	0	
498	42	STD		558	00	0	
499	09	09		559	00	0	
500	01	1					
501	03	3					
502	69	DP			LABELS		
503	01	01					
504	69	DP		046	11	A	
505	05	05		107	23	LNX	
506	61	GTD		293	99	PRT	
507	04	04		421	10	E'	
508	55	55		495	16	A'	
509	76	LBL		510	17	B'	
510	17	B'	(33)	524	19	D'	
511	01	1					
512	04	4					
513	69	DP					
514	01	01					
515	69	DP					
516	05	05					
517	01	1					
518	42	STD					
519	09	09					
520	61	GTD					
521	04	04					
522	55	55					
523	76	LBL					
524	19	D'	(34)				
525	22	INV					
526	86	STF					
527	01	01					
528	22	INV					
529	86	STF					
530	02	02					
531	22	INV					
532	86	STF					
533	03	03					
534	22	INV					
535	86	STF					
536	04	04					
537	68	NDP					
538	68	NDP					
539	68	NDP					

TABLE 3-14a

Program 32 RPN Listing

PRP ""

```
01+LBL "ITR"
02 "VOL?"
03 PROMPT
04 RCL 07    (1)
05 -
06 STO IND 14
07 1
08 ST+ 14
09 "PH?"
10 PROMPT
11 STO IND 14
12 1
13 ST+ 14
14 RCL 14
15 STO 59    (2)
16 GTO "ITR"

17+LBL F    (3)
18 RCL 59
19 2
20 -
21 STO 14

22+LBL "FA" (4)
23 CLΣ
24 RCL 14
25 STO 16

26+LBL "FB" (5)
27 RCL IND 16
28 X=0?
29 GTO "FC"
30 RCL 21
31 X()Y
32 X<=Y?
33 GTO "PK"
34 2
35 ST- 16
36 GTO "FB"

37+LBL "PK" (6)
38 STO 17
39 1
40 ST+ 16
41 RCL 15
42 RCL 17
43 +
44 1/X
45 RCL 17
46 *
47 RCL 18
48 *
49 RCL IND 16
50 STO 19
51 CHS
52 10↑X
53 RCL 09    (7)
54 *
55 +
56 RCL IND 16
57 14
58 -         (8)
59 10↑X
60 RCL 09
```

```
61 *
62 -
63 ENTER↑
64 ENTER↑
65 RCL 17
66 RCL 15
67 +
68 1/X
69 RCL 18
70 *
71 RCL 21
72 *
73 X()Y
74 -
75 /
76 LOG
77 RCL 09
78 *
79 ST- 19
80 RCL 21    (9)
81 RCL 17
82 -
83 10
84 LN
85 *
86 1/X
87 ST+ 01
88 X↑2
89 ST+ 02
90 RCL 19
91 FS? 03    (10)
92 GTO 04
93 FIX 3
94 PRX
95 FIX 9

96+LBL 04    (11)
97 ST+ 03
98 X↑2
99 ST+ 04
100 1
101 ST+ 06
102 3
103 ST- 16
104 GTO "FB"

105+LBL "FC" (12)
106 SF 03
107 TONE 5
108 RCL 21
109 PSE
110 MEAN
111 X()Y
112 PSE
113 SDEV
114 X()Y
115 PRX
116 STO 11
117 CHS
118 RCL 12
119 +
120 PSE
121 RCL 12    (13)
122 1
```

```
123 X=Y?
124 GTO 10
125 RDN
126 RCL 11
127 FS? 04
128 GTO "PR"
129 X<=Y?
130 GTO 03
131 FS? 01
132 GTO 11
133 GTO 02

134+LBL 11
135 FS? 02
136 GTO 12
137 SF 02

138+LBL 03
139 SF 01

140+LBL 10
141 RCL 11
142 STO 12
143 FS? 02
144 GTO 13
145 RCL 08
146 ST+ 21
147 GTO "FA"

148+LBL 13
149 RCL 10
150 ST- 21
151 GTO "FA"

152+LBL 12
153 SF 04
154 CF 03
155 RCL 10
156 ST+ 21
157 GTO "FA"

158+LBL 02
159 -1
160 ST* 08
161 ST* 10
162 GTO 03

163+LBL "PR" (14)
164 TONE 9
165 RCL 06
166 2
167 -
168 STO 16
169 1/X
170 1
171 X()Y
172 -
173 1/X
174 STO 00
175 1.21336
176 *
177 .00864
178 +
179 RCL 00
```

129

TABLE 3-15a

Program 32 RPN Listing and Example

```
180 X↑2          223 "CI= "         265 ASTO 02         ID= CRESOL  (16)
181 1.00559      224 ARCL 17        266 CLA             BASE TITRANT
182 *            225 AVIEW          267 ARCL 01         N= 0.100
183 +            226 "PKA= "        268 ARCL 02         VI= 47.500
184 RCL 00       227 MEAN           269 PROMPT          XS= 0.000
185 3            228 ARCL Y                              INIT VE= 5.000
186 Y↑X          229 AVIEW          270*LBL A                   10.157  ***
187 -.26934      230 "CI(PKA)= "    271 CHS                     10.151  ***
188 *            231 ARCL 19        272 STO 09                  10.151  ***
189 +            232 AVIEW          273 "ACID "                 10.149  ***
190 STO 00       233 ADV           274 ARCL 01                 10.164  ***
191 BEEP         234 ADV           275 ARCL 02                 10.163  ***
192 RCL 04       235 ADV           276 AVIEW                   10.150  ***
193 RCL 03       236 ADV           277 GTO 01                  10.161  ***
194 RCL 06       237 ADV                                       10.151  ***
195 /            238 STOP          278*LBL E                  0.005995849  ***
196 X↑2                            279 STO 09               0.005981532  ***
197 RCL 06       239*LBL E  (15)   280 "BASE "              0.006257831  ***
198 *            240 ΣREG 01       281 ARCL 01              0.006216247  ***
199 -            241 CF 01         282 ARCL 02              0.006181237  ***
200 RCL 16       242 CF 02         283 AVIEW               0.006147755  ***
201 /            243 CF 03                                  0.006115838  ***
202 RCL 02       244 CF 04         284*LBL 01             0.006085396  ***
203 RCL 01       245 CLRG          285 "N= "              0.006056664  ***
204 X↑2          246 .01           286 PROMPT             0.006036703  ***
205 RCL 06       247 STO 08        287 STO 18             0.006015266  ***
206 /            248 .001          288 ARCL 18            0.005997568  ***
207 -            249 STO 10        289 PRA                0.005981532  ***
208 /            250 22            290 "VI= "             0.005970231  ***
209 SQRT         251 STO 14        291 PROMPT             0.005958529  ***
210 RCL 00       252 1            292 STO 15             0.005953792  ***
211 *            253 STO 12        293 ARCL 15            0.005950738  ***
212 STO 17       254 "ID?"         294 PRA                0.005950300  ***
213 RCL 02       255 AON           295 "XS= "             0.005952555  ***
214 RCL 06       256 PROMPT        296 PROMPT                    10.160  ***
215 /            257 AOFF          297 STO 07                    10.153  ***
216 SQRT         258 ASTO 05       298 ARCL 07                   10.153  ***
217 *            259 "ID= "        299 PRA                       10.150  ***
218 STO 19       260 ARCL 05       300 "INIT VE= "               10.165  ***
219 FIX 4        261 AVIEW         301 PROMPT                    10.164  ***
220 "VE= "       262 "TITRANT"     302 STO 21                    10.151  ***
221 ARCL 21      263 ASTO 01       303 ARCL 21                   10.162  ***
222 AVIEW        264 ASHF          304 PRA                       10.152  ***
                                   305 GTO "ITR"           0.005950300  ***
                                   306 .END.             VE= 5.0050
                                                         CI= 0.0217
                                                         PKA= 10.1565
                                                         CI(PKA)= 0.0077
```

(3)   Excess titrant is subtracted from each data point before storage. The prompting messages do not ask for this excess. It must be placed manually in R_{07} if subtraction is desired. There is room for only 13 data points in this program but no extra points are needed for V_e calculation as in program 31. Actual values stored are printed. The pointer is changed (R_{14}) and operation waits for entry of the pH.

(4)   Each pH value is stored in its appropriate data storage register and then printed. The pointer is changed and stored permanently at R_{49}. Permanent storage of the location of the last point entered is required for the iteration process.

(5)   Program operation returns to the beginning in preparation for a new data point. The call for the last data point is left on the printout even though there are no more data to be entered. This program uses direct addresses with a minimum of labels. As a matter of fact, besides the user-defined labels A′, D′, and E′, only LNX and PRT are used as labels. The label LNX calculates pH and label PRT prints the results.

(6)   Label A begins the calculation by calling the pointer from R_{49}. It is necessary to subtract 2 to find the location of the X value of the last data point because R_{49} was poised for storage of a data point that was never entered.

(7)   This short sequence PGM 01 SBR CLR clears the statistics registers R_1–R_6 and leaves the others untouched.

(8)   The data point is examined to see if it is zero. A zero at R_{20} is the signal that all data points in the set have been examined. In this event, program operation moves to step 201 to calculate the standard deviation SD of the set of pK_a values (see decision map).

(9)   The program finds a new V_e. If the V_e being tested is less than the volume of titrant for some of the last data points, these points must be excluded from the calculation—a new pointer is needed for this new data set. The new pointer is in R_{16}.

(10)   An X value that is less than V_e' is stored in R_{17} for the pK_a calculation.

(11)   The sequence beginning at LBL LNX calculates the pK_a according to Eqs. (3-2c) or (3-3c) depending on whether an acid or a base is being titrated. R_{09} contains the factor $+1$ or -1 that provides the correct form of the equation. For curve B, Fig. 3-7, place minus signs at steps 110 and 162.

(12)   The pK_a and (pK_a)² are placed in the appropriate statistics registers R_{01} and R_{02}, respectively. Op 23 places n in R_{03}.

(13)   The pointer is changed for the calculation of the pK_a for the next data point.

(14)   A value of 14 is correct for pK_ω at 25°C. Other values may be substituted by replacing the NOPs.

(15)   The sequence (steps 177–192) calculates ω (see Section II). The values of ω and ω^2 are placed in the statistics registers R_{03} and R_{04}, respectively.

(16)   When all pK_as have been calculated, flag 3 is set and the V_e', as well as the mean pK_a, its standard deviation SD_1, and the difference from

the previous standard deviation are printed. Flag 3 is set unless this is the first or the last calculation of SD. If it is the first or the last, the individual pK_a values are printed (step 293).

(17)   The pause instructions at steps 205, 207, and 218 may be eliminated by replacing them with NOP (as may PRT at step 210) or replaced by PRT to allow progress of the iteration to be studied.

(18)   When calculation begins, R_{12} contains 1 and the increment in R_{08} is added to V'_e (step 262).

(19)   On all but the first iteration, SD_1 is compared to SD_0. If $SD_1 < SD_0$, program operation moves to step 250.

(20)   The manner in which flags 1 and 2 control the sign and the size of the volume increments in R_{08} and R_{10} is best explained by examining the decision map, Fig. 3-8. The sequence from steps 235–291 involves the lower half of the decision map and is explained in Section VI.B.(4).

(21)   The subroutine LBL PRT performs the statistics calculations and prints the summary statement.

(22)   Two degrees of freedom are needed for calculation of the t value, which is stored in R_{16}. The sequence from steps 294–347 calculates the Student's t value for the 95% confidence interval.

(23)   The 95% confidence intervals are calculated for both V_e and the pK_a. For pK_a the confidence interval is calculated without regard for the fact that pK_a is the logarithm of the reciprocal of the acid dissociation constant K. The confidence intervals provide a measure of the precision of the determination but are not to be taken too literally. The Henderson–Hasselbach equation has been used in an approximate form for the derivation of the calculation.

(24)   During the design of this program, labels were first added, then removed. The label pK_a could be added by replacing PRT with Op 05 and removing one ADV. Other labels would require redesign.

(25)   Label E' initializes calculator operation.

(26)   The pointer is set up with 22 as the number where X_1 will be stored.

(27)   The register R_{12} is provided with a suitably large number, 1, so that the first SD_1 value will be less.

(28)   The program asks for data. The first item is the nature of the titrant (see notes 32 and 33).

(29)   The program asks for normality of the titrant. The answer is stored in R_{18} and printed.

(30)   The initial volume (V_i) is required next. This is stored in R_{15} and printed.

(31)   The program requires an estimated equivalent volume to begin calculations.

(32)   Label A' is pressed if the titrant is an acid. The letter A is printed and the factor (-1) is stored in R_{09}.

(33)   Label B' is pressed for titration with base. The letter B is printed and the factor (1) is stored in R_{09}.

(34)   Label D' will begin calculations again after an error has been corrected. Before D' is pressed, 1 is stored in R_{12} and V'_e in R_{21}. LBL D' resets the flags and sends program operation to LBL A.

(35)   This short sequence changes the sign of $\Delta 1$ and $\Delta 2$ in R_{08} and R_{10}, respectively.

2. Program 32 RPN

(1)   Excess volume in 07 is subtracted from titrant volume entries. Data entries are not printed.

(2)   There is more room for data storage in the HP-41C calculator using this program. The pointer is saved in R_{60}.

(3)   LBL F initiates the calculation.

(4)   LBL "FA" clears the statistics registers.

(5)   LBL "FB" ensures that titrant volumes used in calculation are less than the V_e being tested. When titrant volume is 0, program operation moves to LBL "FC."

(6)   LBL "PK" calculates a pK_a value.

(7) R_{09} contains $+1$ or -1, depending on whether the titrant is a base or an acid, respectively.

(8)   The value of K_w at 25.0°C is 14. A value of 14.167 would be entered here if the temperature of the titrated solution is 20.0°C.

(9)   The sequence 80–89, inclusive, calculates ω and ω^2 (see Section II).

(10)   Flag 3 is set after the results of the first iteration are individually printed. It is reset for the printing of the final iteration.

(11) ΣpK and Σ(pK)^2 are stored in the statistics registers.

(12)   A tone announces each iteration, the results of which may be momentarily viewed. Print commands may be substituted for pauses, if desired.

(13)   The changes in the standard deviation (SD) control the sign and magnitude of the increments added to the equivalent volume (see Fig. 3-8).

(14)   LBL "PR" performs the statistics calculations and prints the summary statement. Two degrees of freedom are needed for the calculation of the t value, which is stored in R_{00}. A beep signal at step 191 indicates that the final results are about to be printed.

(15) LBL "E" initiates the calculator for data entry. See notes (28)–(33) of program 32. Program 32 RPN does not use label D'. When an error is corrected after the first iteration, program operation may be started again by placing V'_e in R_{21}, and 1 in R_{12}, clearing flags 1–4 and pressing LBL "F."

(16) The data input and TI-59 version of this example are found in Table 3-3.

VII. PROGRAM 33—PARTITION COEFFICIENT

A. Design

Program 33 is designed to be used while work is in progress to convert raw titration data to partition coefficients and present the raw data and the results in acceptable form for direct entry into a laboratory notebook. Much of the length of this program is taken up simply to record the various labels for the experimental data. Initialization provides an appropriate label to the calculation: partition coefficient or pK_a calculation. Subsequent prompting by the printout also serves as a label for the constants: titrant, N (normality), pK_a, V_o (volume of octanol), V_w (volume of water), and V_e (equivalent volume).

Calculations involved in this program are relatively simple. However, a different equation is required depending on whether the titrant is an acid or a base and whether the ion formed during the titration is protonated or not. These options are handled effectively with flags and operator-controlled labels (Fig. 3-9 and Table 3-17). The program appears complex because there are so many options, not only as to titrant and ion formed but what result is to be calculated. Four different calculations are anticipated: only the pK_a and its standard deviation are required; only the partition coefficient and the apparent partition coefficient are needed; the pK_a is known but pK' and the partition coefficients are required or the calculation is to involve the pK_a and its SD, the pK' and its SD, as well as log P and log P'.

Program 33 uses only 14 memories and these mostly to store a value prior to its recall for printing (Table 3-17). However, it uses five flags and seven "if" statements. Twenty-nine labels are used, nine of these being user-defined. The program would be faster with direct labeling but errors would then be much more difficult to unravel. Operation of the program is best described by reference to Fig. 3-9, the decision map.

The flags are used as follows:

Flag 1   To distinguish the two equations depending on whether the titrant is an acid or a base (A' or B').

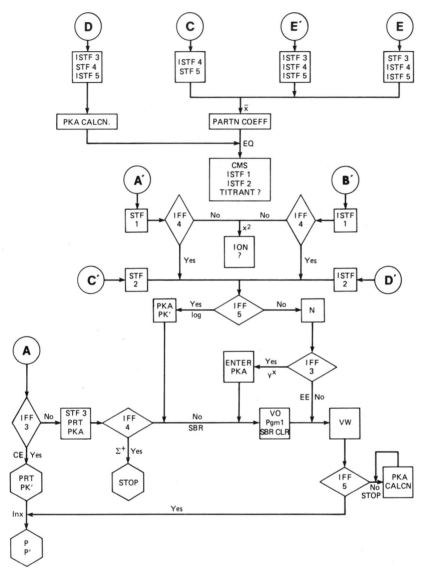

Fig. 3-9 Decision map for program 33, partition coefficient. C, D, E or E′ selects the type of calculation. A′ (or B′) selects the titrant. C′ (or D′) selects the ion formed. Individual p*K* values are calculated automatically when data are entered. A provides the summary statement.

TABLE 3-17

Program 33 Partition Coefficient

Labels			Registers			
001	68	NOP	0		10	pK_a
006	78	Σ+	1	pK	11	V_e
009	35	1/X	2	$(pK)^2$	12	V_w
020	87	IFF	3	n	13	ml
024	89	π	4		14	pH
029	16	A'	5		15	pK'
046	33	X²	6		16	SD
065	17	B'	7		17	V_o
087	18	C'	8	N	18	P
105	19	D'	9		91	P'
124	11	A				
164	71	SBR				
195	52	EE				
245	42	STO				
280	34	ΓX				
320	88	DMS				
346	45	YX				
366	24	CE				
395	23	LNX				
413	77	GE				
461	69	OP				
502	53	(				
524	15	E				
538	14	D				
580	10	E'				
593	79	X̄				
638	67	EQ				
676	13	C				
687	28	LOG				

Flag 2   To distinguish the two equations depending on whether the ion formed is protonated or not (C' or D').

Flag 3   To handle the program when pK_a is known and pK' and $\log P$ are to be calculated (E).

Flag 4   To handle the program when only a pK_a is required (D).

Flag 5   To handle the program when only $\log P$ is required (C).

Basically, the program accepts individual titration points (volume of titrant and corresponding pH of the titrated solution) and calculates the corresponding pK. When several pKs are calculated, Label A is pressed to obtain their mean and SD. When both pK_a and pK' have been calculated, $\log P$ and $\log P'$ are computed and printed.

It may be desirable to change the value of pK_w to correspond to a temperature other than 20°C. The results are in fix decimal 2 format. Other formats are possible.

There are many ways in which programs similar to program 33 may be written for any programmable calculator. Such programs will first calculate average values of pK_a and pK'_a. Equation (3-2c) or (3-3c) is used, depending on whether the titrant is an acid or a base, respectively. The partition coefficient P is then calculated with Eq. (3-1a) if the ion is protonated or with Eq. (3-2a) if the ion is nonprotonated. The apparent partition coefficient P' is calculated with Eq. (3-1b) if the titrant is an acid or with Eq. (3-2b) if the titrant is a base.

B. Program Listing

Program 33, partition coefficient, requires a partitioning of 799.19 (2 Op 17). It requires two cards (three-card sides) for recording. The program listing is provided in Tables 3-18, 3-19, and 3-20 for programs 331, 332, and 333, respectively. The label locations are presented in Table 3-17, and the decision map, Fig. 3-9, should be consulted in tracing operation of this program. Printer symbol designations and numbers for program notes are provided in the margins of the program listings. The general comments of Chapter 1, Section VI apply.

C. Program Notes

Programs 331, 332, 333

(1) The first four labels control the calculation of pK_a. They are placed at the beginning of the program to minimize the time lag in locating these subroutines.

(2) Label A' and label B' set or reset flag 1 to control the calculation of pK_a when the titrant is an acid or a base, respectively.

(3) If flag 4 is not set (see decision map), the program must calculate the partition coefficient. Label X^2 asks the nature of the ion, i.e., whether it is protonated or not. Program operation stops here because a user-defined label (C' or D') will supply the answer.

(4) Label C' and D' set or reset flag 2 to control the calculation of the log P when the ion is nonprotonated or protonated, respectively. Operation then moves to label (to set up the calculation (see program step 502).

(5) Label A calculates the mean pK_a from the series of values already calculated individually. If flag 3 is set, then pK_a has already been entered and label A calculates pK' and then log P and prints the result. If flag 3 is not set, the mean pK_a is calculated and printed. Now if the flag 4 is set (see note 3), the program operation is complete. Note that label $\Sigma+$ at step 006 is one step—R/S. If flag 4 is not set, then the statistics registers are cleared (subroutine SBR, step 164) and preparations are made for calculation of pK' (subroutine EE, step 195) (see decision map). Now if

TABLE 3-18

Program 331 Listing

000	76	LBL		060	01	01		120	02	02		180	03	3	M
001	68	NOP	(1)	061	69	OP		121	61	GTO		181	00	0	
002	94	+/-		062	05	05		122	53	(		182	02	2	L
003	61	GTO		063	91	R/S		123	76	LBL		183	07	7	
004	69	OP		064	76	LBL		124	11	A	(5)	184	69	OP	
005	76	LBL		065	17	B'	(2)	125	87	IFF		185	04	04	
006	78	Σ+		066	69	OP		126	03	03		186	43	RCL	
007	91	R/S		067	00	00		127	24	CE		187	17	17	
008	76	LBL		068	03	3	O	128	86	STF		188	69	OP	
009	35	1/X		069	02	2		129	03	03		189	06	06	(10)
010	75	-		070	02	2	H	130	69	OP		190	36	PGM	
011	43	RCL		071	03	3		131	00	00		191	01	01	
012	14	14		072	02	2		132	03	3	P	192	71	SBR	
013	94	+/-		073	00	0	-	133	03	3		193	25	CLR	(11)
014	22	INV		074	69	OP		134	02	2	K	194	76	LBL	
015	28	LOG		075	04	04		135	06	6		195	52	EE	
016	85	+		076	69	OP		136	01	1	A	196	00	0	
017	61	GTO		077	05	05		137	03	3		197	69	OP	
018	34	ΓX		078	22	INV		138	69	OP		198	04	04	(12)
019	76	LBL		079	86	STF		139	04	04		199	04	4	V
020	87	IFF		080	01	01		140	79	Σ̄		200	02	2	
021	61	GTO		081	87	IFF		141	42	STO		201	04	4	W
022	88	DMS		082	04	04		142	10	10		202	03	3	
023	76	LBL		083	53	(		143	69	OP		203	69	OP	
024	89	π		084	61	GTO		144	06	06	(6)	204	01	01	
025	94	+/-		085	33	X²	(3)	145	22	INV		205	69	OP	
026	61	GTO		086	76	LBL		146	79	Σ̄		206	05	05	(13)
027	77	GE		087	18	C'	(4)	147	42	STO		207	03	3	M
028	76	LBL		088	69	OP		148	16	16		208	00	0	
029	16	A'	(2)	089	00	00		149	03	3	S	209	02	2	L
030	69	OP		090	03	3	O	150	06	6		210	07	7	
031	00	00		091	02	2		151	01	1	D	211	69	OP	
032	02	2	H	092	03	3	R	152	06	6		212	04	04	
033	03	3	+	093	05	5		153	69	OP		213	91	R/S	
034	04	4		094	02	2		154	04	04		214	42	STO	
035	07	7		095	00	0	-	155	43	RCL		215	12	12	
036	69	OP		096	69	OP		156	16	16		216	69	OP	
037	04	04		097	04	04		157	69	OP		217	06	06	(14)
038	69	OP		098	69	OP		158	06	06	(7)	218	87	IFF	
039	05	05		099	05	05		159	98	ADV		219	05	05	
040	86	STF		100	86	STF		160	87	IFF		220	23	LNX	(15)
041	01	01		101	02	02		161	04	04		221	00	0	
042	87	IFF		102	61	GTO		162	78	Σ+	(8)	222	69	OP	
043	04	04		103	53	(		163	76	LBL		223	04	04	
044	53	(		104	76	LBL		164	71	SBR		224	04	4	V
045	76	LBL		105	19	D'	(4)	165	69	OP		225	02	2	
046	33	X²	(3)	106	69	OP		166	00	00		226	01	1	E
047	69	OP		107	00	00		167	04	4	V	227	07	7	
048	00	00		108	03	3	N	168	02	2		228	69	OP	
049	02	2	I	109	01	1		169	03	3	O	229	01	01	
050	04	4	O	110	02	2	H	170	02	2		230	69	OP	
051	03	3	N	111	03	3		171	69	OP		231	05	05	(16)
052	02	2		112	04	4	+	172	01	01		232	03	3	
053	03	3		113	07	7		173	69	OP		233	00	0	
054	01	1		114	69	OP		174	05	05	(9)	234	02	2	
055	02	2	-	115	04	04		175	91	R/S		235	07	7	
056	00	0		116	69	OP		176	42	STO		236	69	OP	
057	00	0		117	05	05		177	17	17		237	04	04	
058	00	0		118	22	INV		178	22	INV		238	91	R/S	
059	69	OP		119	86	STF		179	58	FIX		239	42	STO	

TABLE 3-19

Program 332 Listing

```
240  11  11         300  08  08         360  42  STO        420  53  (
241  69  OP         301  65  ×          361  10  10         421  43  RCL
242  06  06  (17)   302  43  RCL        362  99  PRT        422  12  12
243  98  ADV        303  11  11         363  61  GTO        423  85  +
244  76  LBL        304  55  ÷          364  71  SBR        424  43  RCL
245  42  STO (18)   305  53  (          365  76  LBL        425  11  11
246  91  R/S        306  43  RCL        366  24  CE         426  55  ÷
247  22  INV        307  12  12         367  03  3  ]P      427  02  2
248  58  FIX (19)   308  85  +          368  03  3          428  54  )
249  42  STO        309  43  RCL        369  02  2  ]K      429  55  ÷
250  13  13         310  13  13         370  06  6          430  43  RCL
251  99  PRT        311  54  )          371  06  6          431  17  17
252  91  R/S        312  54  )          372  05  5          432  95  =
253  42  STO        313  95  =          373  69  OP         433  42  STO
254  14  14         314  28  LOG        374  04  04         434  18  18
255  99  PRT        315  87  IFF        375  79  ?  (22)    435  69  OP
256  43  RCL        316  01  01         376  42  STO        436  06  06
257  12  12         317  87  IFF        377  15  15         437  02  2  ]L
258  85  +          318  94  +/-        378  69  OP         438  07  7
259  43  RCL        319  76  LBL        379  06  06         439  02  2  ]G
260  13  13         320  88  DMS        380  22  INV        440  02  2
261  95  =          321  85  +          381  79  x̄         441  03  3
262  35  1/X        322  43  RCL        382  42  STO        442  03  3  ]P
263  65  ×          323  14  14         383  16  16         443  69  OP
264  43  RCL        324  95  =          384  03  3  ]S      444  04  04
265  13  13         325  42  STO        385  06  6          445  43  RCL
266  65  ×          326  15  15  (20)   386  01  1  ]D      446  18  18
267  43  RCL        327  78  Σ+         387  06  6          447  28  LOG
268  08  08         328  03  3  ]P      388  69  OP         448  69  OP
269  87  IFF        329  03  3          389  04  04         449  06  06
270  01  01         330  02  2  ]K      390  43  RCL        450  43  RCL
271  35  1/X        331  06  6          391  16  16         451  10  10
272  85  +          332  01  1  ]A      392  69  OP         452  75  -
273  43  RCL        333  03  3          393  06  06  (22)   453  07  7
274  14  14         334  69  OP         394  76  LBL        454  93  .  (24)
275  94  +/-        335  04  04         395  23  LNX (23)   455  04  4
276  22  INV        336  43  RCL        396  98  ADV        456  95  =
277  28  LOG        337  15  15         397  58  FIX        457  87  IFF
278  75  -          338  58  FIX        398  02  02         458  02  02
279  76  LBL        339  02  02  (21)   399  03  3  ]P      459  68  NOP
280  34  ΓX         340  69  OP         400  03  3          460  76  LBL
281  53  (          341  06  06         401  69  OP         461  69  OP
282  43  RCL        342  98  ADV        402  04  04         462  22  INV
283  14  14         343  61  GTO        403  43  RCL        463  28  LOG
284  75  -          344  42  STO        404  10  10         464  85  +
285  01  1          345  76  LBL        405  75  -          465  01  1
286  04  4          346  45  Y×         406  43  RCL        466  95  =
287  93  .          347  69  OP         407  15  15         467  35  1/X
288  01  1          348  00  00         408  95  =          468  65  ×
289  06  6          349  03  3  ]P      409  87  IFF        469  43  RCL
290  07  7          350  03  3          410  02  02         470  18  18
291  54  )          351  02  2  ]K      411  89  π          471  95  =
292  22  INV        352  06  6          412  76  LBL        472  42  STO
293  28  LOG        353  01  1  ]A      413  77  GE         473  19  19
294  95  =          354  03  3          414  22  INV        474  03  3  ]P
295  55  ÷          355  69  OP         415  28  LOG        475  03  3
296  53  (          356  01  01         416  75  -          476  06  6
297  94  +/-        357  69  OP         417  01  1          477  05  5
298  85  +          358  05  05         418  95  =          478  69  OP
299  43  RCL        359  91  R/S        419  65  ×          479  04  04
```

TABLE 3-20
Program 333 Listing

480	43	RCL		540	86	STF		600	03	3	R	660	03	3	N
481	19	19		541	05	05		601	05	5		661	01	1	
482	69	OP		542	22	INV		602	03	3	T	662	03	3	T
483	06	06		543	58	FIX		603	07	7		663	07	7	
484	02	2	L	544	86	STF		604	02	2	I	664	02	2	-
485	07	7		545	04	04		605	04	4		665	00	0	
486	02	2	G	546	22	INV		606	69	OP		666	00	0	
487	02	2		547	86	STF		607	01	01		667	00	0	
488	06	6		548	03	03		608	03	3	T	668	00	0	
489	05	5		549	69	OP		609	07	7		669	00	0	
490	69	OP		550	00	00		610	02	2	I	670	69	OP	
491	04	04		551	03	3	P	611	04	4		671	02	02	
492	43	RCL		552	03	3		612	03	3	O	672	69	OP	
493	19	19		553	02	2	K	613	02	2		673	05	05	
494	28	LOG		554	06	6		614	03	3	N	674	91	R/S	
495	69	OP	(25)	555	01	1	A	615	01	1		675	76	LBL	
496	06	06		556	03	3		616	00	0		676	13	C	(37)
497	98	ADV		557	00	0		617	00	0		677	22	INV	
498	98	ADV		558	00	0		618	69	OP		678	58	FIX	
499	98	ADV		559	01	1	C	619	02	02		679	86	STF	
500	91	R/S		560	05	5		620	01	1	C	680	05	05	
501	76	LBL		561	69	OP		621	05	5		681	22	INV	
502	53	(	(26)	562	01	01		622	03	3	O	682	86	STF	
503	87	IFF		563	01	1	A	623	02	2		683	04	04	
504	05	05		564	03	3		624	01	1	E	684	61	GTO	
505	28	LOG		565	02	2	L	625	07	7		685	79	$\bar{x}$	(38)
506	69	OP		566	07	7		626	02	2	F	686	76	LBL	
507	00	00	(27)	567	01	1	C	627	01	1		687	28	LOG	
508	03	3		568	05	5		628	02	2	F	688	69	OP	
509	01	1	N	569	03	3	N	629	01	1		689	00	00	
510	69	OP		570	01	1		630	69	OP		690	03	3	P
511	01	01		571	00	0		631	03	03		691	03	3	
512	69	OP		572	00	0		632	69	OP		692	02	2	K
513	05	05		573	69	OP		633	05	05		693	06	6	
514	91	R/S		574	02	02		634	87	IFF		694	01	1	A
515	42	STO		575	69	OP		635	05	05		695	03	3	
516	08	08		576	05	05		636	33	x^2	(35)	696	69	OP	
517	99	PRT		577	61	GTO		637	76	LBL		697	01	01	
518	87	IFF		578	67	EQ	(33)	638	67	EQ	(36)	698	69	OP	
519	03	03		579	76	LBL		639	47	CMS		699	05	05	
520	45	Y×	(28)	580	10	E'	(34)	640	22	INV		700	91	R/S	
521	61	GTO		581	22	INV		641	86	STF		701	42	STO	
522	52	EE	(29)	582	86	STF		642	01	01		702	10	10	
523	76	LBL		583	05	05		643	22	INV		703	99	PRT	
524	15	E	(30)	584	22	INV		644	86	STF		704	03	3	P
525	22	INV		585	58	FIX		645	02	02		705	03	3	
526	86	STF		586	22	INV		646	69	OP		706	02	2	K
527	05	05		587	86	STF		647	00	00		707	06	6	
528	22	INV		588	03	03		648	03	3	T	708	06	6	.
529	58	FIX		589	22	INV		649	07	7		709	05	5	
530	22	INV		590	86	STF		650	02	2	I	710	69	OP	
531	86	STF		591	04	04		651	04	4		711	01	01	
532	04	04		592	76	LBL		652	03	3	T	712	69	OP	
533	86	STF		593	79	$\bar{x}$		653	07	7		713	05	05	
534	03	03		594	69	OP		654	03	3	R	714	91	R/S	
535	61	GTO		595	00	00		655	05	5		715	42	STO	
536	79	$\bar{x}$	(31)	596	03	3	P	656	01	1	A	716	15	15	
537	76	LBL		597	03	3		657	03	3		717	99	PRT	
538	14	D	(32)	598	01	1	A	658	69	OP		718	61	GTO	(39)
539	22	INV		599	03	3		659	01	01		719	71	SBR	

flag 5 is set (label C, step 676) operation moves to LNX (step 395) to calculate log P. If flag 5 is not set, pK' must be calculated so program operation moves to label STO (step 245).

(6) pK_a is printed and labeled.

(7)   SD is printed and labeled.

(8)   Label $\Sigma+$ is at step 006. If flag 4 is set, operation is complete.

(9)   Volume of octanol is requested.

(10)   The volume is printed and labeled in milliliter units.

(11)   The statistics registers are cleared.

(12)   The fourth quarter of the print registers are cleared by 0 OP 04.

(13)   The volume of water (initial volume of solution) is requested.

(14)   The volume is printed and labeled in milliliter units.

(15)   Label LNX is at step 395.

(16)   The equivalent volume of titrant (V_e) is requested.

(17)   The volume is printed and labeled in milliliter units.

(18)   Label STO begins the pK_a calculation. Operation stops until a titrant volume is entered.

(19)   The entered volume is printed as entered without any fix decimal. Fix decimal had been removed at step 178 but it is possible to arrive at label STO from step 344, where fix decimal format was used to print the last pK_a. The calculation of pK_a is straightforward (see programs 31 and 32).

(20)   The calculated pK_a is summed in the statistics registers.

(21)   Two places of decimals are enough for most standard equipment. Other values could be used, however.

(22)   The mean pK_a and the SD are printed and labeled.

(23)   Label LNX calculates the partition coefficient (P), its logarithm (log P), the apparent partition coefficient P' and its logarithm (log P'). V_w is the solution volume ($V_w + \frac{1}{2}V_e$).

(24)   The apparent partition coefficient is calculated automatically at pH 7.4. Other values could be used.

(25)   LP' seems appropriate as a three-symbol designation for log P'. There is no room in the program for a complete label.

(26)   The program comes here after C' or D' are pressed. If flag 5 is set, log P is to be calculated directly and operation moves to LOG (step 687).

(27)   The normality of the titrant is requested. The answer is printed.

(28)   If flag 3 is set, the pK_a is to be calculated. Label y^x is at step 346.

(29)   Label EE is at step 195.

(30)   Label E is pressed if pK_a is known and pK' and the partition coefficients are to be calculated.

(31)   Label $\bar{X}$ is at step 593.

(32)   Label D is pressed if only the pK_a is to be calculated. An appropriate label, PKA CALCN, is printed.

(33)   Label EQ is at step 638.

(34)   Label E' is pressed for the complete calculation of pK_a, pK', and the partition coefficients. The label PARTITION COEFF is printed.

(35)   Label X^2 is at step 46.

(36)   Label EQ prepares for entry of the nature of the titrant.

(37)   Label C is pressed if only the partition coefficients are calculated.

(38)   Label $\bar{X}$ is at step 593.

(39)   There is room for 80 more steps in the program but this would require another bank for program storage.

VIII.   PROGRAM 34—TITRATION CURVE

A.   Design

This program is designed to provide the pH of an aqueous solution of a monoprotic acid or base or the corresponding salt under a variety of circumstances. The titrant may be a strong base or acid, and octanol may be present or absent. Three user-defined labels are used for initiation: E' to begin a new calculation, A to repeat the calculation with a new volume of octanol, a new initial solution volume and/or a new volume of titrant. Label B is used to repeat the same calculation with a new titrant volume (see Fig. 3-10).

The program calculates a partition factor f and then three coefficients B, C, and D. It then solves the appropriate equation (3-5), (3-6), (3-7), or (3-8) for $[H^+]$ by the method of successive approximations (Newton–Raphson technique), using Eq. (3-9). The initial guess of $[H^+]$ is 1 (pH = 0). This unreasonable estimate covers all possibilities and the initial guess is changed by small increments $\frac{1}{8}[H^+]$ until a reasonably close value is reached. The reasonableness is judged when the absolute difference between two successive pH results begin to diminish (i.e., until convergence begins). When this happens the last result for $[H^+]$ becomes the next guess, i.e., $[H^+]_n$. Each succeeding result then becomes the guess for the next calculation until the difference between successive pH values is less than 0.001. Naturally, this assures that the result is correct to the two places of decimals that are printed. It would be equally true for three places of decimals. Fix decimal format is defined at step 175 and may be changed if desired.

This program provides a pH value for all reasonable values of X includ-

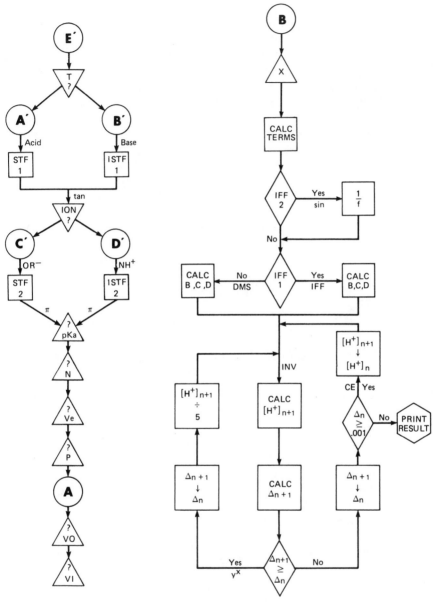

Fig. 3-10 Decision map for program 34, titration curve. E' initializes the program. A' (or B') selects the titrant, and C' (or D') selects the ion formed. A asks for the remaining constants. B performs the iterative calculation via two major loops: first by large intervals until the range is found, and then with finer tuning until an acceptable result is obtained and printed.

ing $X \leq 0$ and $X \geq V_e$. Occasionally, however, the division of $[H^+]_n$ by the factor 5 fails to cause convergence. When this happens, a smaller factor, such as 2, has been found to be successful. The change is made at step 415. This change, however, slows down the iteration process. A pause introduced at step 146 will enable the operator to follow operation. Occasionally the printout is in engineering format with 00 after the pH. This may be prevented by placing INV EE at step 176 before OP 06.

The operation of program 34 is explained in the program notes and in Fig. 3-10. Register contents are listed in Table 3-21.

B. Program Listing

Program 34, titration curve, requires a partitioning of 479.59. It may be recorded on both sides of a single card. Program listing is provided in Tables 3-22 and 3-23 for programs 341 and 342, respectively. Some of the register contents are noted in the margins of the tables together with printer symbol designations and numbers for program notes. Label locations are provided at the bottom of Table 3-23. The general comments of Chapter 1, Section VI, apply.

Table 3-22a is a printout on the HP-41C calculator for the corresponding program in reverse polish notation (RPN) suitable for use with the HP-67, HP-97, or HP-41C calculators. It is designed for use without a printer but may easily be modified by supplying appropriate print instructions.

Program 34 as listed provides pH values at individual titration points. It may be made automatic to provide pH values at specified titrant increments. The increments are summed in R_{18}. The following steps provide for volume increments of 0.01 ml but other suitable increments may be used.

(1) Press GTO 320; replace R/S STO 19 with RCL 18.
(2) Press GTO 180; press Ins 7 times; insert 0.01 SUM 18; GTO B.
(3) Press GTO 146, press Ins, insert Pause.

TABLE 3-21

Program 34

Register contents			
00 9	10 $[H^+]_n$	05 $[X]$	15 D
01 K_w	11 $[H^+]_{n+1}$	06 N	16 V_e
02 K_a	12 f or $1/f$	07 V_o	17 V_i
03 $[V_e]$	13 B	08 P	18 X
04 V_w	14 C	09 0.001	19 n/V_w

TABLE 3-22

Program 341 Listing

#				#				#				#			
000	76	LBL		060	65	×		120	65	×		180	91	R/S	
001	87	IFF	(1)	061	43	RCL		121	43	RCL		181	76	LBL	
002	43	RCL		062	12	12	$f(1/f)$	122	10	10	$[H^+]_n$	182	16	A'	
003	02	02	K_a	063	65	×		123	33	X²		183	86	STF	
004	65	×		064	53	(		124	85	+		184	01	01	(9)
005	43	RCL		065	43	RCL		125	02	2		185	69	OP	
006	12	12	$f(1/f)$	066	03	03	$[V_e]$	126	65	×		186	00	00	
007	85	+		067	75	-		127	43	RCL		187	01	1	A
008	43	RCL		068	43	RCL		128	10	10	$[H^+]_n$	188	03	3	
009	03	03	$[V_e]$	069	05	05	$[x]$	129	65	×		189	69	OP	
010	75	-		070	54	)		130	43	RCL		190	03	03	
011	43	RCL		071	85	+		131	13	13	B	191	69	OP	
012	05	05	$[x]$	072	43	RCL		132	85	+		192	05	05	
013	95	=		073	01	01	K_w	133	43	RCL		193	61	GTO	
014	42	STO		074	95	=		134	14	14	C	194	30	TAN	(10)
015	13	13	B	075	94	+/-		135	54	)		195	76	LBL	
016	43	RCL		076	42	STO		136	95	=		196	17	B'	
017	02	02	K_a	077	14	14	C	137	42	STO		197	22	INV	
018	65	×		078	43	RCL		138	11	11	$[H^+]_{n+1}$	198	86	STF	
019	43	RCL		079	02	02	K_w	139	43	RCL		199	01	01	(11)
020	12	12	$f(1/f)$	080	65	×		140	00	00		200	69	OP	
021	65	×		081	43	RCL		141	32	X:T		201	00	00	
022	43	RCL		082	12	12	$f(1/f)$	142	43	RCL		202	01	1	B
023	05	05	$[x]$	083	65	×		143	11	11	$[H^+]_{n+1}$	203	04	4	
024	85	+		084	43	RCL		144	35	1/X		204	69	OP	
025	43	RCL		085	01	01	K_w	145	28	LOG		205	03	03	
026	01	01	K_w	086	95	=		146	75	-		206	69	OP	
027	95	=		087	94	+/-		147	43	RCL		207	05	05	
028	94	+/-		088	42	STO		148	10	10		208	76	LBL	
029	42	STO		089	15	15	D	149	35	1/X		209	30	TAN	(10)
030	14	14	C	090	76	LBL		150	28	LOG		210	69	OP	
031	43	RCL		091	22	INV	(3)	151	95	=		211	00	00	
032	02	02	K_a	092	43	RCL		152	50	I×I		212	02	2	I
033	65	×		093	10	10	$[H^+]_n$	153	77	GE	(6)	213	04	4	
034	43	RCL		094	75	-		154	45	Y×		214	03	3	O
035	12	12	$f(1/f)$	095	53	(		155	42	STO		215	02	2	
036	65	×		096	43	RCL		156	00	00		216	03	3	N
037	43	RCL		097	10	10	$[H^+]_n$	157	43	RCL		217	01	1	
038	01	01	K_w	098	45	Y×		158	09	09	(7)	218	69	OP	
039	94	+/-		099	03	3		159	32	X:T		219	01	01	
040	95	=		100	85	+		160	43	RCL		220	69	OP	
041	42	STO		101	43	RCL		161	00	00		221	05	05	
042	15	15	D	102	10	10	$[H^+]_n$	162	77	GE		222	91	R/S	
043	61	GTO		103	33	X²		163	24	CE	(8)	223	76	LBL	
044	22	INV		104	65	×		164	03	3	P	224	89	π	
045	76	LBL		105	43	RCL		165	03	3		225	69	OP	
046	88	DMS		106	13	13	B	166	02	2	H	226	00	00	
047	43	RCL		107	85	+		167	03	3		227	03	3	P
048	02	02	K_a	108	43	RCL		168	69	OP		228	03	3	
049	65	×		109	10	10	$[H^+]_n$	169	04	04		229	02	2	K
050	43	RCL		110	65	×		170	43	RCL		230	06	6	
051	12	12	$f(1/f)$	111	43	RCL		171	11	11	$[H^+]_{n+1}$	231	01	1	A
052	85	+		112	14	14	C	172	35	1/X		232	03	3	
053	43	RCL		113	85	+		173	28	LOG		233	69	OP	
054	05	05	$[x]$	114	43	RCL		174	58	FIX		234	01	01	
055	95	=		115	15	15	D	175	02	02		235	69	OP	
056	42	STO		116	54	)		176	69	OP		236	05	05	
057	13	13	B	117	55	÷		177	06	06	pH	237	91	R/S	
058	43	RCL		118	53	(	(4)	178	22	INV		238	99	PRT	
059	02	02	K_a	119	03	3		179	58	FIX		239	94	+/-	

TABLE 3-23

Program 342 Listing

240	22	INV		300	91	R/S		360	01	1		420	22	INV	
241	28	LOG		301	42	STO		361	95	=		421	76	LBL	
242	42	STO		302	17	17		362	87	IFF		422	18	C'	(21)
243	02	02	(13)	303	99	PRT		363	02	02	(17)	423	86	STF	
244	03	3	N	304	76	LBL		364	38	SIN		424	02	02	
245	01	1		305	12	B	(15)	365	76	LBL		425	69	OP	
246	69	OP		306	69	OP		366	42	STO		426	00	00	
247	01	01		307	00	00		367	42	STO		427	03	3	
248	69	OP		308	04	4	X	368	12	12	$f(1/f)$	428	02	2	
249	05	05		309	04	4		369	87	IFF		429	03	3	
250	91	R/S		310	69	OP		370	01	01	(9)	430	05	5	
251	42	STO		311	01	01		371	87	IFF	(1)	431	02	2	
252	06	06		312	09	9	(16)	372	61	GTO	(2)	432	00	0	
253	99	PRT		313	42	STO		373	88	DMS		433	69	OP	
254	04	4	V	314	00	00		374	76	LBL	(18)	434	03	03	
255	02	2		315	01	1		375	10	E'		435	69	OP	
256	01	1	E	316	42	STO		376	69	OP		436	05	05	
257	07	7		317	10	10		377	00	00		437	61	GTO	
258	69	OP		318	69	OP		378	03	3	T	438	89	π	
259	01	01		319	05	05		379	07	7		439	76	LBL	
260	69	OP		320	91	R/S		380	07	7	?	440	19	D'	(22)
261	05	05		321	42	STO		381	01	1		441	22	INV	
262	91	R/S		322	18	18	x	382	69	OP		442	86	STF	
263	42	STO		323	99	PRT		383	01	01		443	02	02	
264	16	16		324	85	+		384	47	CMS		444	69	OP	
265	99	PRT		325	43	RCL		385	01	1	(19)	445	00	00	
266	03	3	P	326	17	17	V_i	386	52	EE		446	03	3	N
267	03	3		327	95	=		387	01	1		447	01	1	
268	69	OP		328	42	STO		388	04	4		448	02	2	H
269	01	01		329	04	04	V_w	389	94	+/-		449	03	3	
270	69	OP		330	35	1/X		390	42	STO	K_w	450	04	4	+
271	05	05		331	65	×		391	01	01		451	07	7	
272	91	R/S		332	43	RCL		392	22	INV		452	69	OP	
273	42	STO		333	06	06	N	393	52	EE		453	03	03	
274	08	08		334	95	=		394	93	.		454	69	OP	
275	99	PRT		335	42	STO		395	00	0		455	05	05	
276	76	LBL		336	19	19	N/V_w	396	00	0		456	61	GTO	
277	11	A	(14)	337	65	×		397	01	1		457	89	π	
278	69	OP		338	43	RCL		398	42	STO		458	76	LBL	
279	00	00		339	16	16	V_e	399	09	09	(20)	459	38	SIN	1/f
280	04	4	V	340	95	=		400	69	OP		460	35	1/X	
281	02	2		341	42	STO		401	05	05		461	61	GTO	
282	03	3	O	342	03	03	$[V_e]$	402	91	R/S		462	42	STO	
283	02	2		343	43	RCL		403	76	LBL	(7)	463	00	0	
284	69	OP		344	19	19	N/V_w	404	24	CE		001	87	IFF	
285	01	01		345	65	×		405	43	RCL		046	88	DMS	
286	69	OP		346	43	RCL		406	11	11		091	22	INV	
287	05	05		347	18	18	x	407	42	STO		182	16	A'	
288	91	R/S		348	95	=		408	10	10		196	17	B'	
289	42	STO		349	42	STO		409	61	GTO		209	30	TAN	
290	07	07		350	05	05	$[x]$	410	22	INV		224	89	π	
291	99	PRT		351	43	RCL		411	76	LBL	(6)	277	11	A	
292	04	4	V	352	07	07	V_o	412	45	YX		305	12	B	
293	02	2		353	65	×		413	42	STO		366	42	STO	
294	02	2	I	354	43	RCL		414	00	00		375	10	E'	
295	04	4		355	08	08	P	415	05	5		404	24	CE	
296	69	OP		356	55	÷		416	22	INV		412	45	YX	
297	01	01		357	43	RCL		417	49	PRD		422	18	C'	
298	69	OP		358	04	04	$[V_w]$	418	10	10		440	19	D'	
299	05	05		359	85	+		419	61	GTO		459	38	SIN	

Bottom right block labels (spelling LABELS): L, A, B, E, L, S

TABLE 3-22a

Program 34 RPN Listing

01♦LBL 67	60 RCL 20	120 STO 09	178♦LBL 15 acid
02♦LBL 10	61 RCL 01	121 RCL 24	179♦LBL a
03♦LBL A (1)	62 *	122 STO 25	180 SF 01
04 STO 06 x	63 RCL 06	123 RCL 09	181 RTN
05 RCL 02 V_i	64 -	124 1/X	
06 +	65 RCL 08	125 LOG	182♦LBL 16 base
07 STO 09 V_w	66 +	126 PSE pH_{n+1}	183♦LBL b
08 1/X	67 STO 21 B	127 RCL 08	184 CF 01
09 RCL 03 N·	68 RCL 20	128 1/X	185 RTN
10 *	69 RCL 01	129 LOG pH_n	
11 STO 24 N/V_w	70 *	130 -	186♦LBL 17 OR^-
12 RCL 06	71 RCL 06	131 ABS Δ_{n+1}	187♦LBL c
13 *	72 *	132 X>Y?	188 SF 02
14 STO 06 $[x]$	73 RCL 00 ·	133 GTO 02	189 RTN
15 RCL 24	74 +	134 STO 24	
16 RCL 07 V_e	75 CHS	135 .001	190♦LBL 18 NH^+
17 *	76 STO 22 C	136 STO 25	191♦LBL d
18 STO 06 $[V_e]$	77 RCL 20	137 RCL 24	192 CF 02
19 RCL 04 V_o	78 RCL 00	138 X>Y?	193 RTN
20 RCL 05 P	79 *	139 GTO 03	
21 *	80 RCL 01	140 RCL 09	194♦LBL 15
22 RCL 09 V_w	81 *	141 1/X	195♦LBL a
23 /	82 CHS	142 LOG	196 STOP
24 1	83 STO 23 D	143 RTN	197 .END.
25 +			
26 FS?C 02 (2)	84♦LBL 07	144♦LBL 02	Labels
27 GTO 04	85 9	145 STO 24	E - Initiate
	86 STO 24 Δn	146 RCL 08	A - New x
28♦LBL 05	87 1	147 5	D - New V_o
29 STO 20 $f(1/f)$	88 STO 08 $[H^+]n$	148 /	
30 FS? 01 (3)		149 STO 08	
31 GTO 06	89♦LBL 01 (4)	150 GTO 01	
32 RCL 01 K_a	90 RCL 08		
33 *	91 3	151♦LBL 03	Registers (7)
34 RCL 06	92 Y↑X H^3	152 RCL 09	20 - $f(1/f)$
35 +	93 RCL 08	153 STO 08	21 - B
36 STO 21 B	94 X↑2 BH^2	154 GTO 01	22 - C
37 RCL 08	95 RCL 21		23 - D
38 RCL 06	96 *	155♦LBL 14	24 - $N/V_w \rightarrow 9 \rightarrow \Delta_n$
39 -	97 +	156♦LBL E (5)	00 - K_w
40 RCL 01	98 RCL 08	157 14-	01 - K_a
41 *	99 RCL 22	158 10↑X	02 - V_i
42 RCL 20	100 * CH	159 STO 00 K_w	03 - N
43 *	101 +	160 RTN	04 - V_o
44 RCL 00	102 RCL 23 D	161 CHS	05 - P
45 +	103 +	162 10↑X	06 - $x \rightarrow [x]$
46 CHS	104 RCL 08	163 STO 01 K_a	07 - V_e
47 STO 22 C	105 X↑2 $3H^2$	164 RTN	08 - $[V_e] \rightarrow 1 \rightarrow [H^+]_n$
48 RCL 20	106 3	165 STO 02 V_i	09 - $V_w \rightarrow [H^+]_{n+1}$
49 RCL 01	107 *	166 RTN	
50 *	108 RCL 08	167 STO 03 N	
51 RCL 00	109 RCL 21	168 RTN	
52 *	110 * 2H	169 STO 07 V_e	
53 CHS	111 *	170 RTN	
54 STO 23 D	112 *	171 STO 05 P	
55 GTO 07	113 +	172 RTN	
	114 RCL 22 C		
56♦LBL 04 (2)	115 +	173♦LBL 13	
57 1/X	116 / $fH/f'H$	174♦LBL D (6)	
58 GTO 05	117 CHS	175 STO 04 V_o	
	118 RCL 08 H_{q+1}	176 RTN	
59♦LBL 06 (3)	119 +	177 GTO 10	

147

The steps should be performed in the order indicated because each insertion alters the position of subsequent steps.

For instructions see Section IV.D.3.

C. Program Notes

1. Programs 341, 342 (See Fig. 3-10)

(1) LBL IFF calculates B, C, and D when the titrant is an acid.

(2) LBL DMS calculates B, C, and D when the titrant is a base.

(3) LBL INV begins the calculation of $[H^+]_{n+1}$. $f[H^+]_n$ is calculated first.

(4) Next $f'[H^+]_n$ is calculated.

(5) R_{00} originally contains a large number 9. Subsequently it has Δ_n.

(6) $\Delta_{n+1} = [pH_{n+1}^- - pH_n]$ is calculated and compared with Δ_n. If Δ_{n+1} is equal or larger than Δ_n, program operation moves to label y^x at step 412. Here Δ_{n+1} becomes Δ_n and a new trial value of $[H^+]_n$ is selected by dividing the previous value by 5 (see decision map, Fig. 3-10).

(7) If Δ_{n+1} is not greater or equal to Δ_n, the value of $[H^+]_n$ is a reasonable value with which to begin the Newton–Raphson calculation. The value of Δ_{n+1} is now tested to see if it meets the final criterion, i.e., that $\Delta_{n+1} - \Delta_n < 0.001$. If not, iteration continues. At label CE (step 404) $[H^+]_{n+1}$ becomes $[H^+]_n$ and program operation moves to label INV.

(8) When the criterion is met that $\Delta_{n+1} - \Delta_n < 0.001$, the pH is calculated, labeled, and the result is printed.

(9) Flag 1 is set if the titrant is an acid.

(10) LBL TAN at step 209 continues the prompting.

(11) Flag 1 is reset if the titrant is a base.

(12) Prompting now asks for the constants.

(13) pK_a is converted to K_a before storing in R_{02}.

(14) Label A is used to avoid reentering the conditions for a repeat calculation.

(15) Label B is used for repeat calculations when only the volume of titrant is changed.

(16) The numbers 9 and 1 are stored in R_{00} and R_{10}, respectively, to initial the calculations.

(17) Flag 2 is set if the ion is nonprotonated (see step 459).

(18) Label E' initializes the program.

(19) The value of K_w (1×10^{-14}) is the value for a temperature of 25°C.

(20) The criterion for stopping the iteration is $pH_{n+1} - pH_n < 0.001$.

(21) Flag 2 is set if the ion is not protonated.

(22) Flag 2 is reset if the ion is protonated.

2. *Program 34 RPN*

(1) LBL A initiates the calculation for a new X.

(2) Flag 2 is set if the ion is nonprotonated. In this case, the factor f is replaced by $1/f$.

(3) Flag 1 is set if the titrant is an acid. In this case, LBL b applies the appropriate equations for the calculation of the coefficients B, C, and D.

(4) LBL 1 begins the calculation of $[H^+]_{n+1}$. $f[H^+]_n$ is calculated first. Refer to programs 341, 342, notes for details.

(5) LBL E enters the constants and raw data.

(6) LBL D is used when the calculation is repeated with a new V_0.

(7) These are HP-41C designations. For HP-67/HP-97, the corresponding data registers are 0–9 for R_{00}–R_{09} and A–E for R_{20}–R_{24}.

REFERENCES

1. Y. C. Martin, "Quantitative Drug Design: A Critical Introduction." Marcel Dekker, New York, 1978.
2. C. Hansch and A. Leo, "Substituent Constants for Correlation Analysis in Chemistry and Biology." Wiley, New York, 1979.
3. R. F. Rekker, "The Hydrophobic Fragmental Constant." Elsevier, New York, 1977.
4. H. Kubinyi, Lymphophilicity and drug activity. *In* "Progress in Drug Research," Vol. 23 (E. Jucker, ed.). Berkhaüser Verlag, Basel, 1978.
5. A. Albert and E. P. Serjeant (a) "Ionization Constants of Acids and Bases." Methuen, London, 1962. (Now out of print.) (b) "The Determination of Ionization Constants, A Laboratory Manual." Chapman and Hall, Ltd., London, 1971. (Out of print.) Available on loan from the National Library of Medicine, Bethesda 14, MD/File Code QD 561 A 333 i (1971).
6. R. F. Cookson, The determination of acidity constants. *Chem. Rev.* **74**, 5, 1974.
7. I. M. Kolthoff and N. H. Furman, "Potentiometric Titrations," pp. 95, 96. Wiley, New York, 1949.
8. J. M. H. Fortuin, Method for the determination of the equivalent point in potentiometric titrations. *Anal. Chem. Acta* **24**, 175, 1961.
9. S. Ebel and A. Seuring, Fully automatic potentiometric titrations. *Angew. Chem. Int. Ed. Engl.* **16**, 157, 1977.
10. D. M. Barry and L. Meites, Titrimetric applications of multiparametric curve-fitting: Part I. Potentiometric titrations of weak bases with strong acids at extreme dilutions. *Anal. Chim. Acta* **68**, 435, 1974.
11. T. N. Briggs and J. E. Stuehr, Simultaneous determination of precise equivalent points and pK values from potentiometric data: Single pK systems. *Anal. Chem.* **46**, 1517, 1974.
12. L. Meites, J. E. Stuehr, and T. N. Briggs, Simultaneous determination of precise equivalence points and pK values from potentiometric data: Single pK systems. *Anal. Chem.* **47**, 1465, 1975.
13. M. Box, "An on-line computer method for the potentiometric titration of mixtures of a strong and a weak acid. *Anal. Chim. Acta* **90**, 61, 1977.

14. E. W. Wentworth, Rigorous least-squares adjustment: Application to some nonlinear equations, I. *J. Chem. Ed.* **42**, 96, 1965.
15. W. E. Gordon, Component discrimination in acid–base titration. *J. Phys. Chem.* **83**, 1365, 1979.
16. J. J. Kaufman, N. M. Semo, and W. S. Koski, Microelectrometric titration measurement of the pK_as and partition and drug distribution coefficients of narcotics and narcotic antagonists and their pH and temperature dependence. *J. Med. Chem.* **18**, 647, 1975.
17. P. Seiler, The simultaneous determination of partition coefficient and acidity constant of a substance. *Eur. J. Med. Chem.* **9**, 663, 1974.
18. Y. C. Martin, Advances in the methodology of quantitative drug design. *In* "Drug Design," Vol. VIII (E. J. Ariens, ed.). Academic Press, New York, 1979.
19. J. Cymerman-Craig and A. A. Deomontis, Lipoid–water partition coefficients of some aromatic bases. *J. Chem. Soc.*, 1619, 1953.
20. R. A. Scherrer and S. M. Howard, Use of distribution coefficients in quantitative structure–activity relationships. *J. Med. Chem.* **20**, 53, 1977.
21. H. P. Williams, Titration curve for a programmable pocket calculator. *J. Chem. Ed.* **54**, 237, 1979.
22. J. E. Barnes and A. J. Waring, "Pocket Programmable Calculators in Biochemistry, p. 22. Wiley, New York, 1980.
23. E. Still and R. Sara, Calculation of potentiometric titration curves. *J. Chem. Ed.* **54**, 348, 1977.
24. G. E. Knudson and D. Nimrod, Exact equation for calculating titration curves for dibasic salts. *J. Chem. Ed.* **54**, 351, 1978.

CHAPTER 4

Correlation Analysis

I. INTRODUCTION

Statistical analysis of the relationship between variables is of increasing importance in chemistry and biology [1–3]. In a review of the chemical aspects of correlation analysis [1], the authors included the Brönsted equation, the Hammett equation, linear free-energy relationships, nucleophilicity, and the correlation of NMR chemical shifts with Hammett σ values. The use of statistical methods to analyse dose–response relationships in pharmacology is well known [2]. One of the most important advances in understanding the effects of organic compounds in biological systems has been made by the study of quantitative structure–activity relationships (QSAR). This application stimulated the design of programs described in this chapter.

Modern calculators have a built-in capacity to perform linear regression analysis of two variables [4]. The equation for the regression line $y = a_0 + a_1x_1$ is quickly provided in terms of the slope or regression coefficient a_1 and the y intercept a_0. The correlation coefficient R is also provided as well as estimates of new values of the dependent variable y for given values of the independent variable x. The capacity for regression analysis is extended for the TI-59 calculator by the applied statistics module to include the regression and correlation coefficients for trivariate data. The module programs allow for the storage and retrieval of the raw data base so that, at any time, all the data used in the regression equation may be examined and changed as desired. As useful as these programs are, they are deficient in two important aspects: The storage of the raw data base is cumbersome for large sets of data, and confidence intervals for the regression coefficients are not provided.

The programs of this chapter are designed to accommodate easily any number of data points, to provide confidence intervals, and to extend the capacity to include three or four independent variables and one dependent variable. In addition, it is possible to enter trivariate data and to analyze it with respect to each pair of variables.

Thus, some of the programs of this chapter are interrelated. Others are long and require a number of cards to complete the calculation. Table 4-1 lists the programs with a brief designation of their respective functions. Program 41 is to be recorded on two sides of one card and is listed as 411 and 412. It is used with two variables, one of which is independent. Programs 421, 422, and 423 provide statistical parameters for three variables, two of which are independent. Program 424 is entered after the operation of 421, 422, and 423 is complete. It provides the correlation matrix. Programs 425 and 426 provide bivariate equations for each pair of variables for trivariate data. It is used after the data have been entered with programs 421, 422, and 423. Finally, programs 427 and 428 provide a bilinear equation for trivariate data. Programs 427 and 428 are used to enter the data and perform initial calculations. Final statistical parameters are then obtained with programs 421, 422, and 423. The design of the program for the bilinear equation differs from the others because calculation of the parameters is an iterative process performed with nonlinear regression techniques.

The program for four variables, three of which are independent, requires three cards and program operation requires three steps. In step 1,

TABLE 4-1

Programs for Correlation Analysis

No.	Function	No.	Function
411	2 Variables	431	4 Variables begin
412		432	
421		433	Complete calculation
422	3 Variables	434	Residuals
423		435	Correlation matrix, deletion
424	3V Correlation matrix	441	5 Variables begin
425	2V Equations	442	
426		443	Complete calculation
427	Bilinear equation	444	Statistical parameters
428		445	Coefficients
		446	Correlation matrix
		447	Delete data, residuals

programs 431 and 432 are entered. The data are entered and calculation is begun. Programs 433 and 434 are then entered and calculation of statistical parameters is completed. Data points may also be deleted. Program 435 provides for the calculation of residuals and data deletion. Naturally, if data are deleted, programs 431 and 432 must be reentered to calculate the new statistical parameters.

The program for five variables requires six steps to use its full capacity. In the first step, data are entered and calculation is begun with programs 441 and 442. Program 443 is entered to complete the calculation. Program 444 provides statistical parameters and program 445 provides the coefficients of the regression equation. The correlation matrix is provided with program 446 and data may be deleted with program 447.

An older text edited by Owen L. Davis [5] was very helpful for the design of the programs of this chapter. The basic concept is to transform the raw data base immediately upon entry into an intermediate data base that consists of sums of the entered data as well as sums of the squares and pairs of cross products. New sets of data may be added or subtracted from this intermediate data base. When regression coefficients are calculated, the intermediate data base is transformed into the components of a matrix. The matrix is then inverted with the help of the master library module. The components of the inverse matrix are used to calculate the statistical parameters: regression coefficients, correlation coefficient, F statistic, and confidence intervals. This process enables the user to observe quickly the effect on the statistical parameters of the addition or subtraction of data points. Data points are printed as they are entered so there is a record of exactly what data are being used in the regression analysis. The results are also printed and appropriately labeled. The programs are especially useful when a set of raw data is examined for the first time in the search for an appropriate regression equation.

II. CALCULATIONS

Let y be the linearly dependent variable and x be the independent variable, and let (x_1, y_1), (x_2, y_2), $\cdots$, (x_n, y_n) be a set of data points. In Fig. 4-1, these points are seen to be scattered about the regression line AB. The equation of the regression line is $\tilde{y} = a_0 + a_1 x$, where $\tilde{y}$ is the true value of y for a given value of x. a_0 is the intercept of the regression line with the y axis, and a_1 is the slope or regression coefficient. The equation for the line is calculated by the method of least squares to minimize the sum of the squares of the deviations $y - \tilde{y}$.

Let the sum of the squares of the deviations be Q. Then $Q = \Sigma[y - (a_0 + a_1 x)]^2$. a_0 and a_1 are chosen so that Q is a minimum. This is done by differential calculus by partially differentiating Q with

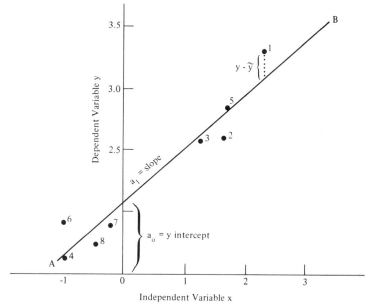

Fig. 4-1   Graph of a dependent variable versus an independent variable for correlation analysis. Data of Table 4-1.

respect to a_0 and a_1, equating each derivative to zero, and solving the resulting equations for a_0 and a_1.

$$dQ/da_1 = -2 \Sigma[y - (a_0 + a_1x)] = 0$$
$$dQ/da_0 = -2 \Sigma x[y - (a_0 + a_1x)] = 0$$

Rearranging,

$$\Sigma y = na_0 + a_1 \Sigma x$$
$$\Sigma xy = a_0 \Sigma x + a_1 \Sigma x^2$$

Whence

$$a_0 = (\Sigma y - a_1 \Sigma x)/n$$
$$a_1 = [\Sigma xy - (\Sigma x \Sigma y)/n]/[\Sigma x^2 - (\Sigma x)^2/n]$$

The expression for a_1 can be expressed differently as

$$a_1 = \Sigma(y - \bar{y})(x - \bar{x})/\Sigma(x - \bar{x})^2$$
$$= C_{1y}/C_{11} = C^{11}C_{1y}$$

where

$$C_{11} = \Sigma(x - \bar{x})^2 = \Sigma x^2 - (\Sigma x)^2/n$$
$$C_{1y} = \Sigma(y - \bar{y})(x - \bar{x}) = \Sigma xy - \Sigma x \Sigma y/n$$

and C^{11} is the inverse of C_{11} and $\bar{x}$ and $\bar{y}$ are the averages $\Sigma x/n$ and $\Sigma y/n$, respectively.

The statistical parameters of the regression equation are obtained from the following equations [5].

$$\text{Correlation coefficient } R = C_{1y}/(C_{11}C_{yy})^{1/2}$$
$$\text{Standard deviation } S = [(C_{yy} - a_1C_{1y})/\varphi]^{1/2}$$

where the number of degrees of freedom, $\varphi = n - 2$ and n is the number of data points.

The F statistic:

$$F = R^2\varphi/(1 - R^2)$$

The confidence interval (CI) for a_0:

$$CI_{a_0} = a_0 \pm ts[n + (\Sigma x)^2/C_{11}]^{1/2}/n$$

and

$$CI_{a_1} = a_1 \pm ts/(C_{11})^{1/2}$$

where t is the Student's t value for φ degrees of freedom and the appropriate probability level (usually 95%). As will be shown below, t can be calculated with excellent accuracy at the 95% probability level using an equation derived by regression analysis from the data for a Student's t table.

Related expressions may be derived for multiple linear regression involving two, three, or four independent variables [5].

The regression equations are

3 variables: $y = a_0 + a_1x_1 + a_2x_2$
4 variables: $y = a_0 + a_1x_1 + a_2x_2 + a_3x_3$
5 variables: $y = a_0 + a_1x_1 + a_2x_2 + a_3x_3 + a_4x_4$

The expressions for a_0 are

3 variables: $a_0 = (\Sigma y - a_1 \Sigma x_1 - a_2 \Sigma x_2)/n$
4 variables: $a_0 = (\Sigma y - a_1 \Sigma x_1 - a_2 \Sigma x_2 - a_3 \Sigma x_3)/n$

and similarly for five variables.

The expressions for the statistical parameters all have similar patterns and it will only be necessary to illustrate the equations for three variables.

$$a_1 = C^{11}C_{1y} + C^{12}C_{2y}$$
$$a_2 = C^{21}C_{1y} + C^{22}C_{2y}$$
$$a_0 = (\Sigma y - a_1 \Sigma x_1 - a_2 \Sigma x_2)/n$$
$$S = [(C_{yy} - a_1C_{1y} - a_2C_{2y})/\varphi]^{1/2}$$
$$R = (1 - S^2\varphi/C_{yy})^{1/2}$$
$$F = R^2\varphi/2(1 - R^2)$$

[for four variables, $F = R^2\varphi/3(1 - R^2)$ and similarly for five variables],

$$CI_{a_0} = (ts/n)(n + C^{11}(\Sigma x_1)^2 + C^{22}(\Sigma x_2)^2 + 2C^{12} \Sigma x_1 \Sigma x_2)^{1/2}$$

(For four variables this becomes

$$CI_{a_0} = (tx/n)[n + C^{11}(\Sigma x_1)^2 + C^{22}(\Sigma x_2)^2 + C^{33}(\Sigma x_3)^2$$
$$+ 2C^{12} \Sigma x_1 \Sigma x_2 + 2C^{13} \Sigma x_1 \Sigma x_3 + 2C^{23} \Sigma x_2 \Sigma x_3]^{1/2}$$

and similarly for five variables),

$$CI_{a_1} = ts(C^{11})^{1/2}$$
$$CI_{a_2} = ts(C^{22})^{1/2}$$

Simple correlation coefficients can be calculated for pairs of variables:

$$R_{1y} = C_{1y}/(C_{11}C_{yy})^{1/2}$$

Partial correlation coefficients may also be obtained:

$$R_{12} = C^{12}/(C^{11}C^{22})^{1/2}$$

The components of the matrix are arranged as follows for four variables, three of which are independent:

$$C_{11} \quad C_{12} \quad C_{13}$$
$$C_{12} \quad C_{22} \quad C_{23}$$
$$C_{13} \quad C_{23} \quad C_{33}$$

When the inverse of the matrix is calculated, its components have a similar pattern:

$$C^{11} \quad C^{12} \quad C^{13}$$
$$C^{12} \quad C^{22} \quad C^{23}$$
$$C^{13} \quad C^{23} \quad C^{33}$$

The sum of the squares due to the regression is

$$a_1 C_{1y} + a_2 C_{2y} + \cdots a_p C_{py} \quad (p = 1, 2, 3, 4)$$
$$SS = C_{yy} - \varphi S^2$$

The residual sum of squares (RSS) = φS^2. The very brief derivations above are simplified and more fully explained in references [4] and [5].

III. EXAMPLES

A. Bivariate Data: $y = a_0 + a_1 x$

Table 4-2 provides data for the correlation of analgesia measured quantitatively in animals with opiate receptor binding measured *in vitro*. Table 4-3 is a printout of the results of the analysis of the data of Table 4-1 using programs 411 and 412. The data are presented graphically in Fig. 4-1.

The first column of Table 4-3 provides the printout of the data entered into the calculator. Each data point is spaced separately. The points are numbered consecutively as they are entered. The second number is the independent (x) variable. When all data points are entered, the statistical parameters are calculated and the results are printed with all digits. With fix 2, the results are reprinted to two places of decimals (top of column 3). The 95% confidence intervals for a_0 and a_1 illustrate the significance of the equation.

The individual data points may be reentered so that calculated values (y_c) of y may be obtained as well as the residuals or differences from the observed values of y.

Point 6 has the largest residual, and it is of interest to see what effect its removal from the data set will have on the statistical parameters. The printout for this calculation is presented in the lower half of the third column. The correlation coefficient improves (from 0.96 to 0.98), the standard deviation is lower (from 0.17 to 0.13), and the F statistic is nearly

TABLE 4-2

Correlation of Analgesia with Opiate Receptor Binding[a]

No.	Compound	log $1/IC_{50}$[b]	log $1/ED_{50}$[c]
1	Levorphanol	2.35	3.27
2	Morphine	1.66	2.58
3	Methadone	1.29	2.58
4	Meperidine	−1.00	1.61
5	I	1.72	2.82
6	II	−1.00	1.90
7	III	−0.22	1.88
8	IV	−0.48	1.71

Log $1/ED_{50}$ = 2.06(0.16) + 0.43(0.12) log $1/IC_{50}$
$\gamma = 0.964$, $S = 0.173$, $n = 8$

[a] From F. H. Clarke, H. Jaggi, and R. A. Lovell, *J. Med. Chem.* **21**, 600, 1978.

[b] IC_{50} is the micromolar concentration of the test compound that reduces specific tritiated naloxone binding by 50% in the absence of sodium.

[c] ED_{50} is a measure of analgetic potency in mice following subcutaneous administration of the test compound.

doubled (from 78 to 144). The 95% confidence intervals for the regression coefficients and the intercept are both somewhat improved.

B. Trivariate Data

1. Equation with Three Variables: $y = a_0 + a_1 x_1 + a_2 x_2$

Table 4-4 provides data for the correlation of cytotoxicity of a series of copper chelates with the substituent constants π and σp. The data are taken from a paper by E. A. Coats *et al.* [6]. The authors used Eqs. (4-1) and (4-2) to describe the bivariate and trivariate correlations of cytotoxicity (expressed quantitatively as pI_{50}) with π, and with π and σp, respectively. Calculator analyses of this data using programs 421, 422, 423, and 424 are presented in Tables 4-4 and 4-5.

$$pI_{50} = 4.86(\pm 0.61) - 0.53(\pm 0.69)\pi \qquad (4\text{-}1)$$
$$n = 7, \quad S = 0.496, \quad R = 0.663$$
$$pI_{50} = 5.05(\pm 0.45) - 0.66(\pm 0.48)\pi - 1.12(\pm 1.04)\sigma p \qquad (4\text{-}2)$$
$$n = 7, \quad S = 0.309, \quad R = 0.909$$

TABLE 4-3

Regression Analysis of Bivariate Data (Table 4-2)

Data input	Residuals		Statistical parameters	
1.	2.35	X	8.00	N
2.35	3.07	YC	0.96	R
3.27	3.27	Y	0.17	S
	0.20	⌂	78.23	F
2.				
1.66	1.66	X	2.06	A0
2.58	2.77	YC	0.16	CI
	2.58	Y	0.43	A1
	-0.19	⌂	0.12	CI
3.				
1.29				
2.58	1.29	X		
	2.61	YC	Data deletion	
	2.58	Y		
4.	-0.03	⌂	7.00	
-1.			-1.00	
1.61			1.90	
	-1.00	X		
	1.64	YC	Revised parameters	
5.	1.61	Y		
1.72	-0.03	⌂	7.00	N
2.82			0.98	R
			0.13	S
	1.72	X	144.51	F
6.	2.80	YC		
-1.	2.82	Y	1.99	A0
1.9	0.02	⌂	0.14	CI
			0.47	A1
7.	-1.00	X	0.10	CI
-0.22	1.64	YC		
1.88	1.90	Y		
	0.26	⌂	Revised residuals	
8.			2.35	X
-0.48	-0.22	X	3.10	YC
1.71	1.97	YC	3.27	Y
	1.88	Y	0.17	⌂
Statistical parameters	-0.09	⌂		
			1.66	X
8.	N	-0.48	X	2.78	YC
.9637272683	R	1.86	YC	2.58	Y
.1732514623	S	1.71	Y	-0.20	⌂
78.23446398	F	-0.15	⌂		
2.062875683	A0			
.1629899845	CI			
.4275450314	A1			
.1183255711	CI			

The first column of Table 4-5 is the calculator printout as the original data are entered into the calculator using programs 421 and 422. Each data point is numbered consecutively and separated with a space. The components of each data point are entered and printed in the sequence x_1, x_2, and y referring in this example to π, σp, and pI_{50}, respectively.

The second column of Table 4-5 is the calculator printout of the results

TABLE 4-4

Correlation of Cytotoxicity[a] with Substituent
Constants[b] of Copper Chelates[c]

No.	Substituent[d]	π	σp	pI_{50}[e]
1	H	0.00	0.00	5.49
2	Br	0.86	0.23	4.05
3	Cl	0.71	0.23	4.37
4	OCH_3	-0.02	-0.27	5.32
5	CH_3	0.56	-0.17	4.52
6	NO_2	-0.28	0.78	4.32
7	C_6H_5	1.96	-0.01	3.93

[a] Inhibition of respiration of Ehrlich ascites tumor cells.

[b] Values of C. Hansch *et al.* [10].

, [c] Data from E. A. Coats *et al.* [6].

[d] Para-substituted [phenylglyoxal bis(4-methyl-3-thiosemicarbazone)] copper chelates.

[e] Logarithm of reciprocal of dose in millimoles per milliliter causing 50% inhibition of respiration.

of the trivariate statistical analysis. The program for trivariate data has two parts (see Table 4-1). The data are entered with programs 421, 422, and 423, and label B is pressed to perform the calculation. The first number of column 2 (2.334 . . .) is the determinant of the matrix used in the calculation. The next four numbers are the elements of the inverse matrix. These numbers are used in the calculation of the statistical parameters but are of no direct interest otherwise. They are printed automatically by the Master Library module program and cannot be left out. The statistical parameters are individually labeled: N, number of data points; R, correlation coefficient; S, standard deviation; F, the F factor or variance ratio; A_0, A_1, and A_2 are the coefficients of the regression equation normally written in lower case as a_0, a_1, and a_2, respectively. The label CI is the 95% confidence interval ($\pm$) of the coefficient that immediately precedes it. The results of column 2 were first printed without a fixed decimal. The decimal was then fixed at 2 and the calculation was repeated (B pressed again). The results are provided by the next set of numbers. These permit us to write the regression equation (4-3):

$$pI_{50} = 5.05(\pm 0.45) - 0.66(\pm 0.48)\pi - 1.11(\pm 1.04)\sigma p \qquad (4\text{-}3)$$
$$n = 7, \quad R = 0.91, \quad S = 0.31, \quad F = 9.57$$

It is interesting to note that our calculator has duplicated the computer results presented in the paper of Coats *et al.* [6].

TABLE 4-5

Statistical Analysis of Trivariate Data (Table 4-3)

Data input	Statistical Parameters		Residuals	
1.	2. 334703863		0. 00	
0.			0. 00	
0.			5. 05	YC
5. 49			5. 49	Y
	. 3113640529		0. 44	⌐
	. 1689415851			
	. 1689415851			
	1. 467289179		0. 86	
	7.	N	0. 23	
	. 9094597981	R	4. 23	YC
	. 3079307028	S	4. 05	Y
	9. 568525762	F	-0. 18	⌐
	5. 054537813	A0	0. 71	
	. 4471092932	CI	0. 23	
	-. 6601141921	A1	4. 33	YC
2.	. 4770851668	CI	4. 37	Y
0. 86	-1. 11383785	A2	0. 04	⌐
0. 23	1. 035665562	CI		
4. 05			-0. 02	
			-0. 27	
3.	2. 33		5. 37	YC
0. 71			5. 32	Y
0. 23			-0. 05	⌐
4. 37				
	0. 31		0. 56	
4.	0. 17		-0. 17	
-0. 02	0. 17		4. 87	YC
-0. 27	1. 47		4. 52	Y
5. 32	7. 00	N	-0. 35	⌐
	0. 91	R		
5.	0. 31	S	-0. 28	
0. 56	9. 57	F	0. 78	
-0. 17			4. 37	YC
4. 52	5. 05	A0	4. 32	Y
	0. 45	CI	-0. 05	⌐
6.	-0. 66	A1		
-0. 28	0. 48	CI	1. 96	
0. 78	-1. 11	A2	-0. 01	
4. 32	1. 04	CI	3. 77	YC
	13.		3. 93	Y
7.	-. 6646163475		0. 16	⌐
1. 96	23.			
-0. 01	-. 4349857712			
3. 93	12.			
	-. 2499450105			

Program 424 is now entered into the calculator without removing the data in the memories. When label A is pressed, the correlation matrix is calculated and printed. For purposes of program economy, the variables x_1, x_2, and y are here referred to simply by the numbers 1, 2, and 3, respectively. Thus, the correlation coefficient for x_1 and y is

−0.6646. . . . Often it is the square of these coefficients that is used and Coats *et al.* [6] provide the value 0.06 for the R^2 between the parameters π and σp for the same seven data points, $[(0.2499 \ldots)^2 = 0.062]$. The usefulness of R is that the sign is retained so that it is possible to conclude that pI_{50} is negatively correlated with both π and σp. This fact is evident, of course, from the sign of the coefficients in regression Eq. (4-3).

Column 3 of Table 4-5 provides the residuals for each point of column 1. As the values of π and σp are entered, the calculated pI_{50} is computed and printed with the label YC. When the found value of pI_{50} is entered as Y, the residual is calculated and printed as Δ. The point with the largest residual is the first one.

2. *Equations with Two Variables* (Table 4-6)

For many reasons, it may be useful to compare the trivariate equation with each of the bivariate equations that could be obtained from the same data set. It is of special interest to know whether the addition of a third variable to a bivariate equation is significant. Programs 425 and 426 were designed for this purpose. With the data still in the calculator, programs 425 and 426 were entered and the statistical parameters for each bivariate equation were calculated in turn.

For x_1 and y, the equation (4-4) may be written as

$$pI_{50} = 4.86(\pm0.61) - 0.53(\pm0.69)\,\pi \qquad (4\text{-}4)$$
$$n = 7, \quad S = 0.495, \quad R = -0.665, \quad F = 3.96$$

The value of F calculated for the addition of the third variable to the equation in π and pI_{50} is 8.9, which is statistically significant. The results are nearly identical to Eq. (4-1) of Coats *et al.* [6].

A similar analysis was made for x_2 and y so that Eq. (4-5) may be written as

$$pI_{50} = 4.66(\pm0.61) - 0.76(\pm1.80)\,\sigma p$$
$$n = 7, \quad S = 0.596, \quad R = -0.435, \quad F = 1.16 \qquad (4\text{-}5)$$

It is apparent from Eq. (4-3) and (4-5) that although σp is not itself correlated with pI_{50} with a high degree of significance, it does contribute in a statistically significant way to Eq. (4-3).

The final set of statistical parameters is available from the calculated data, but does not contribute much to the problem of Coats *et al.* [6] except to demonstrate that π and σp are nearly independent parameters.

Table 4-6 also illustrates the subtraction of data from the three variable data set and the further manipulation of the bivariate data entered with the

TABLE 4-6

Bivariate Regression Equations from Trivariate Data (Table 4-5)

Bivariate parameters		Data deletion		Bivariate residuals	
	X1 Y				
7.	N	6.		0.00	X
-.6646163475	R	0.		4.67	YC
.4949371129	S	0.		5.49	Y
3.955997402	F	5.49		0.82	⌂
8.917071369	F-3V				
				0.86	X
4.859397368	AO	1.98		4.33	YC
.6082243556	CI			4.05	Y
-.5318684893	A1			-0.28	⌂
.6875649116	CI				
		0.36		0.71	X
		0.24		4.39	YC
		0.24		4.37	Y
		1.56		-0.02	⌂
	X2Y	6.00	N		
7.	N	0.95	R		
-.4349857712	R	0.20	S	-0.02	X
.5964519843	S	13.84	F	4.68	YC
1.166844885	F			5.32	Y
14.75922936	F-3V	4.89	AO	0.64	⌂
		0.39	CI		
4.656711276	AO	-0.55	A1	0.56	X
0.614165565	CI	0.38	CI	4.45	YC
-.7556695358	A1	-0.96	A2	4.52	Y
1.798716812	CI	0.79	CI	0.07	⌂
			X1 Y	-0.28	X
		6.	N	4.79	YC
		-.6430000339	R	4.32	Y
	X1X2	.4211364405	S	-0.47	⌂
7.	N	2.819526856	F		
-.2499450105	R	14.99312586	F-3V	1.96	X
.3691961304	S			3.88	YC
.3331769404	F	4.672702234	AO	3.93	Y
18.65810918	F-3V	.6362387507	CI	0.05	⌂
		-.4026948292	A1		
0.175196457	AO	.6658809956	CI		
.4537022435	CI				
-0.115138575	A1				
.5128859771	CI				

trivariate program. For column 2 of Table 4-6, programs 421 and 422 were reentered to replace 425 and 426, and the first point, $\pi = 0$, $\sigma p = 0$, $pI_{50} = 5.49$, is removed with the help of label A'. The statistical parameters of the trivariate equation are printed when label B is pressed. Note that R, S, and F are improved and all of the 95% confidence intervals are smaller.

Programs 425 and 426 were now reentered and the statistical parameters of the bivariate equation in π and pI_{50} were determined. This bivariate equation is hardly an improvement over the bivariate equation with 7 points. Now if programs 411 and 412 are entered, the bivariate data set may be used to calculate the residuals. These are presented in column 3 of Table 4-6.

The data handling illustrated in this example could, of course, be repeated with any number of data points. A small published set was selected to illustrate the use of the calculator programs because the data are quickly entered and manipulated.

3. The Parabolic Model: $y = a_0 + a_1 x_1 + a_2 x_1^2$

Table 4-7 provides data from the work of P. Timmermans *et al.* [7] which illustrate the correlation between the apparent partition coefficient (log P') and penetration of a drug to the central nervous system. The penetration to the brain is expressed by the authors as the log of the ratio of the brain concentration to an intravenously administered dose. The data are supplied for a series of compounds related to the antihypertensive agent, clonidine. The equations of Timmermans *et al.* [7] [Eq. (4-6); (4-7)]

TABLE 4-7

Lipophilicity versus Penetration to Brain[a]

No.	Substitution[b]	log P' [c]	log $\dfrac{C_{brain}}{C_{iv}}$ [d]
1	2,6-di-Cl	0.62	0.262
2	2,6-di-Br	1.21	0.333
3	2,6-di-F	−0.16	−0.200
4	2,4-di-Cl	0.29	0.163
5	2,5-di-Cl	0.65	0.379
6	2-Me,4-Cl	−1.06	−0.866
7	2-Cl,4-Me	−0.48	−0.654
8	2,4-di-Me	−1.66	−1.300
9	H	−1.92	−1.886
10	2-Me	−1.82	−1.390
11	2,4,6-tri-Cl	1.47	0.475
12	2,4,6-tri-Br	2.24	0.583
13	2,6-di-Cl,4-Br	1.97	0.405
14	2,6-di-Me,4-Br	−0.28	−0.283

[a] P. Timmermans *et al.* [7].

[b] Substituents on the benzene ring of clonidine.

[c] P' is the apparent partition coefficient determined in the octanol buffer (pH 7.4) system.

[d] Ratio of drug concentration in the brain to the iv-administered dose.

describe the correlation in quantitative terms as a linear and a quadratic equation, respectively:

$$\log (C_{brain}/C_{iv}) = -0.327 + 0.555 \log P' \qquad (4\text{-}6)$$
$$n = 14, \quad R = 0.947, \quad S = 0.269, \quad F = 104.12$$
$$\log (C_{brain}/C_{iv}) = -0.094 - 0.133(\log P')^2 + 0.574 \log P' \qquad (4\text{-}7)$$
$$n = 14, \quad R = 0.987, \quad S = 0.139, \quad F = 211.83$$

Table 4-8 is a printout of the data of Table 4-7 as it was entered into the calculator and of the calculated statistical parameters obtained with programs 421, 422, and 423. This program has a stop (R/S) command at position 018, which separates the entry of x_1 from that of x_2. The stop follows the calculation of x_1^2. In this instance, the stop command is replaced with NOP (nonoperative), which allows the calculator to print x_1^2 and enter the result as x_2 with only a single operation. When all the data are entered, label B is pressed to calculate the statistical parameters. These are the same as those provided by Timmermans *et al.* in the quadratic equation (4-7). However, our calculator program supplies the 95% confidence intervals for the coefficients. These were not provided by Timmermans *et al.* [7]. We are also able to determine quickly that although $\log P'$ correlates well with $\log (C_{brain}/C_{iv})$, there is a very poor correlation of $(\log P')^2$ with this parameter. Use of programs 425 and 426 provides the parameters of the linear equation (4-7). In addition, we note that the F factor for the addition of $(\log P')^2$ to the bivariate equation is 33.9. This is the same result as was reported by Timmermans *et al.* [7] and shows that the $(\log P')^2$ term makes a significant contribution to Eq. (4-7).

Since Eq. (4-7) is a quadratic equation in $\log P'$, it implies that there is an optimum value of $\log P'$ for penetration of a clonidinelike compound to the brain. This optimum value is found by equating the first derivative of Eq. (4-7) to zero:

$$2(-1.33 \log P') + 0.574 = 0$$
$$\log P' = 2.16$$

This result could be very useful for the further design of clonidinelike compounds.

4. The Bilinear Model: $y = a_0 + a_1 \log P + a_2 \log(\beta P + 1)$

a. Minimum data. Since this program is quite complex, it will be illustrated first with a minimum of data. Table 4-9 provides partition coefficient data for a series of aliphatic carboxylic acids that have been studied for their ability to diffuse through a toad bladder. The data plotted in Fig. 4-2 are entered using programs 427 and 428. Printout of the data entry

TABLE 4-8

Derivation of Quadratic Equation (Data of Table 4-7)

Data input		Parameters Quadratic equation	
1.	8.	909.3418398	
0.62	-1.66		
0.3844	2.7556		
0.262	-1.3		
		.0412837569	
		-.0038265216	
2.	9.	-.0038265216	
1.21	-1.92	.0269921831	
1.4641	3.6864	14.	N
0.333	-1.886	.9872685493	R
		0.13920739	S
3.	10.	211.8843313	F
-0.16	-1.82		
0.0256	3.3124	-.0937091646	A0
-0.2	-1.39	.1203488041	CI
		.5736670689	A1
4.	11.	.0622851332	CI
0.29	1.47	-.1332240542	A2
0.0841	2.1609	.0503632557	CI
0.163	0.475		
5.	12.	13.	
0.65	2.24	.9469183174	
0.4225	5.0176	23.	
0.379	0.583	-.1689791087	
		12.	
6.	13.	.1146292789	
-1.06	1.97		
1.1236	3.8809		
-0.866	0.405		
7.	14.	Linear equation	
-0.48	-0.28		X1 Y
0.2304	0.0784	14.	N
-0.654	-0.283	.9469183174	R
		.2693686132	S
		104.1151357	F
		33.93147185	F-3V
		-.3266153809	A0
		.1571904268	CI
		.5547806853	A1
		0.118518252	CI

is provided in Table 4-10, top of column 1. When label A is pressed, calculation begins. In this case, an initial value of −1 was selected for log β, and the first data printed reflect this. Numbers marked a and b are unavoidable printouts of matrix computations. The β factor, R^2, and F factor are printed after each iteration although they are not labeled. Iteration continues and the printout is quite long. However, the F value for

TABLE 4-9

Correlation of Partition Coefficient with Absorption
through the Bladder[a]

No.	Compound	Log P	Log K
1	Propionic acid	0.38	2.76
2	Butyric acid	0.88	3.20
3	Valeric acid	1.38	3.53
4	Hexanoic acid	1.88	4.26
5	Heptanoic acid	2.38	4.25
6	Octanoic acid	2.88	4.31

[a] Data from H. Kubinyi, *Arzneim. Forsch.* **29**, 1067, 1979.

$\log \beta = -4$ is less than that for $\log \beta = -3$, so the iteration is succeeding. The final result is $\log \beta = -2.65$.

Table 4-11, first column, provides the printout at the end of the calculation. Programs 421, 422, and 423 are now entered with the appropriate changes in degrees of freedom as described in the instructions. The result is provided in the bottom of column 1 of Table 4-11. The derived equation is

$$\log K = 2.35(\pm 0.86) + 1.00(\pm 0.73) \log P - 2.15(\pm 4.11)\log(\beta P + 1)$$
$$n = 6, \quad R = 0.986, \quad S = 0.175, \quad F = 22.61, \quad \log \beta = -2.65$$

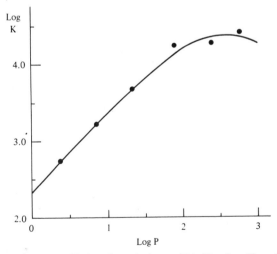

Fig. 4-2 Correlation of diffusion through the toad bladder ($\log K$) and partition coefficient ($\log P$). Black circles are data of Table 4-9. Curve is derived from coefficients of the bilinear equation (Table 4-11).

TABLE 4-10

Derivation of Bilinear Equation (Data of Table 4-9)

0.38				
2.76	.7149574174	.7422818104	.8221438998	.9709193138
	-44.70647919	-15.92792256	-6.868411504	-4.480947078
0.88	-44.70647919	-15.92792256	-6.868411504	-4.480947078
3.2	4109.224623	493.8555384	79.47652373	27.04781291
	-4.	-3.5	-3.	-2.61
1.38	.9694806432	.9699507002	.9708882505	.9713510214
3.53	21.1773935	21.51909712	22.23359453	22.60350559
1.88	.0016488016	.0131623666	.0750436885	.1580583838
4.26				
2.38	.7182942956	.7523305224	.8503770134	.9655776371
4.25	-36.04790401	-13.1943447	-6.020870023	-4.516643714
	-36.04790401	-13.1943447	-6.020870023	-4.516643714
2.88	2653.442327	332.3870346	58.29937318	27.6796453
.4.31	-3.9	-3.4	-2.9	-2.62
	.9695426378	.9701028509	.9710873332	.9713559843
	21.22185609	21.63200347	22.39127784	22.60753743
.3105849072 [a]	.0025393203	.0192978355	0.100033634	.1544122931
7.790169685	.7224689411	.7647713843	.8846923992	.9603340068
-10.32061684 [b]	-29.17082444	-11.02549548	-5.356831244	-4.553372607
-10.32061684	-29.17082444	-11.02549548	-5.356831244	-4.553372607
14.08632518	1722.901972	226.7093633	43.73529008	28.33323638
-1. $\mathrm{Log}\,\beta$	-3.8	-3.3	-2.8	-2.63
.9436094915 R^2	.9696183575	.9702766583	.9712513918	.9713595691
11.15565387 F	21.27640858	21.76239512	22.52286171	22.61045056
.4164268943	.0038846233	.0278638147	.1302389136	0.150813091
1.576327281	.7276841159	.7801253585	.9262382402	.9551867652
-3.762922352	-23.70903507	-9.305993262	-4.841084667	-4.591153237
-3.762922352	-23.70903507	-9.305993262	-4.841084667	-4.591153237
10.50604574	1126.235334	157.013677	33.59211067	29.00941803
-2.	-3.7	-3.2	-2.7	-2.64
0.966786873	.9697101194	.9704699142	.9713497997	.9713618207
19.40571818	21.34288418	21.90917922	22.60251328	22.61228065
.0550477021	.0058947213	.0395423199	.1654881335	.1472614891
.8221438998	.7341878511	.7990083973	.9763607247	.9501342837
-6.868411504	-19.37171071	-7.944412533	-4.446263757	-4.630005629
-6.868411504	-19.37171071	-7.944412533	-4.446263757	-4.630005629
79.47652373	742.1894527	110.6409541	26.43694086	29.70905718
-3.	-3.6	-3.1	-2.6	-2.65
.9708882505	0.969820227	.9706771287	0.971344635	.9713627837
22.23359453	21.42318357	22.06871484	22.59831939	22.61306349
.0010646777	.0088588659	.0550477021	.1617506012	.1437581493

TABLE 4-11

Three Equations in Log P—Data of Tables 4-10 and 4-11

Bilinear equation	Linear equation	X1 Y	Quadratic equation	
.9451749626	6.	N	10.20833333	
-4.669950368	.9482972413	R		
-4.669950368	.2326165862	S		
30.43305734	35.70919283	F	4.783257143	
-2.66	-3.823714295	F-3V	-1.397142857	
.9713625024			-1.397142857	
22.61283474	2.635081905	A0	.4285714286	
	.5682089713	CI	6.	N
.1472614891	.6645714286	A1	.9788512583	R
	.3087879428	CI	.1731308343	S
	Data reentered		34.3421106	F
.9501342837	1.			
-4.630005629	0.38		2.186195429	A0
-4.630005629	0.1444		.8478736426	CI
29.70905718	2.76		1.423685714	A1
-2.65			1.205399228	CI
.9713627837			-.2328571429	A2
22.61306349	2.		.3608117331	CI
	0.88			
.1472614891	0.7744			
	3.2			
	3.			
.9501342837	1.38			
-4.630005629	1.9044			
-4.630005629	3.53		13.	
29.70905718			.9482972413	
6. N	4.		23.	
0.985577386 R	1.88		.8723177994	
.1754028353 S	3.5344		12.	
22.61306349 F	4.26		.9758146705	
2.346927855 A0	5.			
0.864411353 CI	2.38			
.9989021182 A1	5.6644			
.7361782075 CI	4.25			
-2.145278078 A2				
4.116564244 CI				
	6.			
13.	2.88			
.9482972413	8.2944			
23.	4.31			
.6947020065				
12.				
.8714542448				

To obtain the optimum log P, one uses the equation

$$\log P_o = \log[a_1/\beta(-a_2 - a_1)]$$

In this case, $\beta = 10^{-2.65} = 0.00224$ and

$$a_1 = 1.00, \quad a_2 = -2.15, \quad \log P_o = 2.59$$

Program 424 is now entered and label A pressed to give the R value for x_1y. The values for x_2y and x_1x_2 are not meaningful in this instance. However, when programs 425 and 426 are entered, the statistical parameters for the bivariate equation in x_1y are obtained (Table 4-11, column 2). The F-3V value again has no meaning. The bivariate equation is

$$\log K = 2.64(\pm 0.57) + 0.66(\pm 0.30) \log P$$
$$n = 6, \quad R = 0.948, \quad S = 0.23, \quad F = 35.71$$

The trivariate equation is derived using programs 421, 422, and 423 (Table 4-11, columns 2 and 3).

$$\log K = 2.19(\pm 0.85) + 1.42(\pm 1.20) \log P - 0.23(\pm 0.36) \log P^2$$
$$n = 6, \quad R = 0.979, \quad S = 0.173, \quad F = 34.34$$

The F for adding the third variable is 4.22 and the optimum $\log P$ in this case is

$$\log P_o = (-a_1/2a_2) = 3.08$$

For this data set, the F value is best for the simple quadratic equation in $\log P$. However, the bilinear equation, as H. Kubinyi [8] points out, has a theoretical basis and does provide a better fit for data that is more clearly bilinear, i.e., the descending portion of the graph is more distinct than in this example. Such a result is clearly illustrated below.

The bilinear equation is designed to fit data for which two linear equations can be fitted. The data of Table 4-12 illustrate this quite clearly when plotted as in Fig. 4-3. The data of Table 4-12 illustrate, in this instance, the

TABLE 4-12

Correlation of Partition Coefficient with Absorption
across the Blood–Brain Barrier[a]

No.	Compound	Log P	Log $1/C$[b]
1	Methanol	−0.66	0.96
2	Ethanol	−0.16	1.49
3	1-Propanol	0.34	1.88
4	1-Butanol	0.88	2.27
5	1-Pentanol	1.40	2.61
6	1-Hexanol	2.03	2.77
7	1-Heptanol	2.53	2.66
8	1-Octanol	3.03	2.27
9	1-Nonanol	3.53	1.86
10	1-Decanol	4.03	1.50

[a] Data from H. Kubinyi, *Arzneim. Forsch.* **29**, 1067, 1979.
[b] C = Concentration causing neurotoxicity.

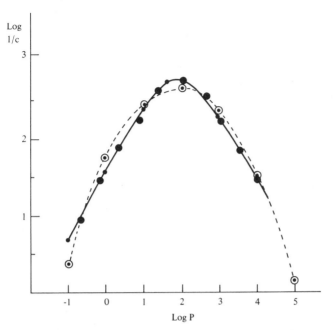

Fig. 4-3 Correlation of Neurotoxicity (log 1/C) and partition coefficient (log P). Large black circles are data of Table 4-12. Small black circles are derived from the bilinear equation and open circles from the quadratic equation, Tables 4-13 and 4-14.

relationship of biological activity and partition coefficient. It is obvious from Fig. 4-3 that there is an optimum value of the partition coefficient for maximum biological effect. Although the data can be fitted to a polynomial equation in log P and $(\log P)^2$, the bilinear equation provides a closer fit to the data. Programs 427, 428, and 421, 422, and 423 illustrate this in the printout of Table 4-13.

The first column of Table 4-13 is the calculator printout as the data are entered with programs 427 and 428. No value for log β is entered so the program begins with log $\beta = 0$. For log $\beta = 0$, the F value is only 5.41 . . . , but it rises quickly as the iteration proceeds. Column 2 (top) provides the second value of R and F as the iteration proceeds. Next is presented the final value of R and F when the optimum log β is reached. Programs 421, 422, and 423 are then entered with the appropriate changes in degrees of freedom (see Section IV). The printout of column 2, Table 4-13, provides R, S, F, the coefficients and the confidence limits. The bilinear equation may be written as

$$\log {}^1/C = 1.58 \pm 0.05 + 0.88 \pm 0.05 \log P - 1.76 \pm 0.10 \log(\beta P + 1)$$
$$n = 10, \quad R = 0.998, \quad S = 0.042, \quad F = 604, \quad \log \beta = -1.95$$

TABLE 4-13

Bilinear and Quadratic Equations for Data of Table 4-12

Bilinear equation		Quadratic equation		
Input	Results	Input	Results	
-0.66	17.32981044	1.	880.5857651	
0.96		-0.66		
		0.4356		
-0.16	.6489232523	0.96	.3389525866	
1.49	-.8961012297		-0.087817349	
	-.8961012297	2.	-0.087817349	
0.34	1.326353227	-0.16	.0261024547	
1.88	-1.	0.0256	10.	N
	.9102901528	1.49	0.988892771	R
0.88	20.29409661		.1005021214	S
2.27		3.	154.9349342	F
	20.51065961	0.34		
1.4		0.1156	1.673922803	A0
2.61		1.88	.1146113575	CI
			1.029822585	A1
	.2477895702	4.	.1384278846	CI
2.03	-0.475575135	0.88	-.2686750997	A2
2.77	-0.475575135	0.7744	.0384144643	CI
	1.107144684	2.27		
2.53	-1.95			
2.66	.9967005577	5.		
	604.163048	1.4		
3.03		1.96		
2.27		2.64		
3.53		6.		
1.86		2.03		
		4.1209		
4.03	20.76101735	2.77		
1.5				
		7.		
	.2477895702	2.53		
5.071700474	-0.475575135	6.4009		
	-0.475575135	2.66		
	1.107144684			
	10.	8.	N	
3.558070871	.9983489158	3.03	R	
-3.991034685	.0417175183	9.1809	S	
-3.991034685	604.163048	2.27	F	
4.532099267				
0.	1.583001483	9.	A0	
.7304021464	.0504395489	3.53	CI	
5.418456688	.8809407127	12.4609	A1	
	.0508341747	1.86	CI	
	-1.756824113		A2	
	0.107452483	10.	CI	
		4.03		
		16.2409		
		1.5		

The corresponding polynomial equation is

$$\log {}^{1}/C = 1.67 \pm 0.11 + 1.03 \pm 0.14 \log P - 0.27 \pm 0.04 \, (\log P)^2$$
$$n = 10, \quad R = 0.988, \quad S = 0.101, \quad F = 155$$

The calculator printout for the derivation of the polynomial equation with programs 421, 422, and 423 is provided in columns 3 and 4 of Table 4-13.

Column 1 of Table 4-14 is a short program that can be used to calculate individual values of the bilinear equation. The coefficients a_0, a_1, and a_2 are stored in R_{01}, R_{02}, and R_{03}, respectively. Log β is stored in R_{04}. The printout of this short program for the bilinear curve of Fig. 4-3 is provided in column 2 of Table 4-14. The corresponding polynomial values are in

TABLE 4-14

Comparison of Calculated Data Using Bilinear and
Quadratic Equations

Program for use of bilinear equation			Data for Fig 4-3	
			Bilinear	Quadratic
000	91	R/S	-1.00	-1.00
001	42	STD	0.70	0.38
002	05	05		
003	99	PRT	0.00	0.00
004	43	RCL	1.57	1.67
005	01	01		
006	85	+		
007	43	RCL	1.00	1.00
008	02	02	2.38	2.44
009	65	×		
010	43	RCL	2.00	2.00
011	05	05	2.77	2.66
012	85	+		
013	43	RCL	3.00	3.00
014	03	03	2.32	2.35
015	65	×		
016	53	(		
017	53	(	4.00	4.00
018	43	RCL	1.50	1.49
019	04	04		
020	65	×	5.00	5.00
021	43	RCL	0.63	0.11
022	05	05		
023	22	INV		
024	28	LOG		
025	85	+		
026	01	1		
027	54	)		
028	28	LOG		
029	54	)		
030	95	=		
031	99	PRT		
032	98	ADV		
033	81	RST		
034	00	0		
035	00	0		

column 3. The first and last points are quite different. Figure 4-3 shows that the bilinear equation is a better fit to the data of Table 4-12.

The bilinear equation and the polynomial provide values of the optimum $\log P$, which are nearly the same.

For the bilinear equation

$$\log P_0 = \log \frac{0.89}{10^{-1.93}(1.76-0.89)} = 1.93$$

For the polynomial equation,

$$\log P_0 = -1.03/(-0.27 \times 2) = 1.91$$

C. Quatrivariate Data: $y = a_0 + a_1x_1 + a_2x_2 + a_3x_3$

The work of Leclerc and his associates (footnote a, Table 4-15) provides an example of the correlation of vasopressive activity with physicochemical parameters for a series of analogs of the drug clonidine. The experimental results of the authors are tabulated in Table 4-15. The first two columns identify the compounds and the next two columns list the steric parameters of substituents in the ortho positions of the phenyl ring. The parameters are entered in column 4 only if there is a substituent at both ortho positions. In this example, the data for one dependent and three independent variables are considered, the square of column 4 data being considered one set of independent data. Addition of a fourth parameter F will be considered in example 5.

Programs 431 and 432 for data entry provide the square of the first point entered and this value remains in the display. If the R/S key is now pressed, the square of the first value will be entered as the second entry. Since, for our purposes, column 4 is the one squared, its values are entered first (Table 4-16). Then the values of column 3 and column 6 are entered as x_3 and y, respectively. When all the data are entered, label B is pressed to begin calculation. The results of the correlation coefficient R and the standard deviation S are printed and labeled following the printout of the determinant and matrix components (Table 4-16, column 3). These are followed by the regression coefficients a_0, a_1, a_2, and a_3 in turn, but these last numbers are not labeled with programs 431 and 432. It would be possible at this point to add data to see if the correlation improves. However, to subtract data, program 435 is required.

To obtain the labeled correlation coefficients with their 95% confidence limits and the matrix of partial correlations, programs 433 and 434 are now entered. Label A is pressed and the results appear as in the remainder of

TABLE 4-15

Vasopressive Activity versus Physicochemical Properties of
2-Aryliminoimidazolines[a]

No.	Substituents[b]	E_{S_2}[c]	$E_{S_{2+6}}$[c]	F[d]	pD_2^*[e]
1	2,6-Di-iPr	−1.71	−3.42	−0.18	5.79
2	H	0	0	0	5.84
3	4-Br	0	0	0.41	6.09
4	2-CF$_3$	0	−2.40	0.35	6.13
5	2-OMe,4-Me	0	−0.25	0.41	6.58
6	2,4-Di-Cl	0	−1.06	1.13	6.71
7	2-Cl,4-Me	0	−1.06	0.71	7.08
8	2-Me,4-Cl	0	−1.24	0.33	7.11
9	2,4,6-Tri-Cl	−1.06	−2.12	1.87	7.13
10	2,6-Di-F	−0.75	−1.50	1.52	7.23
11	2,6-Di-Br	−1.24	−2.48	1.56	7.28
12	2,6-Di-Cl,4-OH	−1.06	−2.12	1.75	7.37
13	2,6-Di-Et	−1.31	−2.62	−0.14	7.61
14	2,6-Di-Cl	−1.06	−2.12	1.48	7.66
15	2,4-Di-Me	0	−1.24	−0.09	7.74
16	2-Cl,6-Me	−1.06	−2.30	0.68	7.74
17	2,6-Di-Me	−1.24	−2.48	−0.12	7.74

[a] Data from R. Rouot et al. J. Med. Chem. **19**, 1049, 1976.
[b] Substituents are on the phenyl ring.
[c] Steric constants for substituents in the 2, and 2 plus 6 positions,
respectively.
[d] Field effect.
[e] Vasopressive Activity is expressed as the log of the reciprocal of
the concentration (moles/kg) of protonated species at pH 7.4.

column 3 of Table 4-16. The corresponding equation obtained by Leclerc
and his associates is

$$pD_2^*$$
$$= -1.57(\pm 0.62)E_{S_{2+6}} - 0.59(\pm 0.20)[E_{S_{2+6}}]^2 - 0.84(\pm 0.58)E_{S_2} + 6.03$$
$$n = 17, \quad r = 0.88, \quad S = 0.36, \quad F = 15$$

Our results are the same for r, S, and F and for the coefficients a_0, a_1,
a_2, and a_3. The first two digits of our confidence intervals are also the same
as those of the authors but our rounded numbers differ in the last digit.
Our program provides a confidence interval for a_0, which is much smaller
than a_0. The partial correlation matrix indicates that x_1 and x_2 are highly
correlated as we would expect, but it also indicates that x_1 and x_2 are not
highly correlated with x_3, if each is adjusted for the omitted variable.

TABLE 4-16

Statistical Analysis of Quatrivariate Data (Table 4-15)

Data input

1.	9.	17.
-3.42	-2.12	-2.48
11.6964	4.4944	6.1504
-1.71	-1.06	-1.24
5.79	7.13	7.74
2.	10.	Parameters
0.	-1.5	429.2669487
0.	2.25	
0.	-0.75	
5.84	7.23	.6567737196
3.	11.	.1794842476
0.	-2.48	-0.087837239
0.	6.1504	.1794842476
0.	-1.24	.0713933409
6.09	7.28	.0715518196
		-0.087837239
4.	12.	.0715518196
-2.4	-2.12	.5791559835
5.76	4.4944	
0.	-1.06	.8809931548 R
6.13	7.37	.3571467185 S
5.	13.	6.031048047
-0.25	-2.62	-1.574435553
0.0625	6.8644	-.5902404895
0.	-1.31	-.8365252728
6.58	7.61	17. N
		.8809931548 R
6.	14.	.3571467185 S
-1.06	-2.12	15.02477607 F
1.1236	4.4944	
0.	-1.06	6.031048047 A0
6.71	7.66	.6283155295 CI
		-1.574435553 A1
7.	15.	.6255580598 CI
-1.06	-1.24	-.5902404895 A2
1.1236	1.5376	.2062474528 CI
0.	0.	-.8365252728 A3
7.08	7.74	.5874318834 CI
8.	16.	R
-1.24	-2.3	.8288760722 12
1.5376	5.29	-.1424206683 13
0.	-1.06	.3518795728 23
7.11	7.74	

Programs 433 and 434 also provide the residuals that are calculated by reentering the data. The program is adjusted at 323–325 by replacing R/S, NOP, NOP with RCL 44 x^2 so that the polynomial term is calculated and printed automatically. Table 4-17 is a printout of the results. We see that

TABLE 4-17

Residuals for Quatrivariate Equation (Table 4-16)

-3.42	X1	-1.24	X1	-1.24	X1
11.6964	X2	1.5376	X2	1.5376	X2
-1.71	X3	0.	X3	0.	X3
5.94	YC	7.08	YC	7.08	YC
5.79	Y	7.11	Y	7.74	Y
-0.15	▵	0.03	▵	0.66	▵
0.	X1	-2.12	X1	-2.3	X1
0.	X2	4.4944	X2	5.29	X2
0.	X3	-1.06	X3	-1.06	X3
6.03	YC	7.60	YC	7.42	YC
5.84	Y	7.13	Y	7.74	Y
-0.19	▵	-0.47	▵	0.32	▵
0.	X1	-1.5	X1	-2.48	X1
0.	X2	2.25	X2	6.1504	X2
0.	X3	-0.75	X3	-1.24	X3
6.03	YC	7.69	YC	7.34	YC
6.09	Y	7.23	Y	7.74	Y
0.06	▵	-0.46	▵	0.40	▵

Correlation matrix

-2.4	X1	-2.48	X1	14.
5.76	X2	6.1504	X2	-.3184814951
0.	X3	-1.24	X3	24.
6.41	YC	7.34	YC	.0760376813
6.13	Y	7.28	Y	34.
-0.28	▵	-0.06	▵	-.3346782489
				13.
-0.25	X1	-2.12	X1	.8289543953
0.0625	X2	4.4944	X2	23.
0.	X3	-1.06	X3	-.8486609946
6.39	YC	7.60	YC	12.
6.58	Y	7.37	Y	-.9487185451
0.19	▵	-0.23	▵	

-1.06	X1	-2.62	X1
1.1236	X2	6.8644	X2
0.	X3	-1.31	X3
7.04	YC	7.20	YC
6.71	Y	7.61	Y
-0.33	▵	0.41	▵
-1.06	X1	-2.12	X1
1.1236	X2	4.4944	X2
0.	X3	-1.06	X3
7.04	YC	7.60	YC
7.08	Y	7.66	Y
0.04	▵	0.06	▵

the largest residual occurs when x_1 is -1.24 and x_2 is 0 (compound 15 in Table 4-13). It would be of interest to subtract this point and observe the effect on the statistical parameters. Program 435 is entered and label A' is pressed and the data of compound 15 are entered. Now programs 431 and

432 are entered again and label B is pressed, then programs 433 and 434 and label A. The printout is in column 1 of Table 4-18. *R, S,* and *F* are all improved, as are the confidence intervals for the coefficients of the regression equation. To confirm our calculations, the deleted point is added back and the original data regenerated as shown in column 3 of Table 4-18. This

TABLE 4-18

Data Deletion and Addition for Quatrivariate Equation (Table 4-16)

No. 15 Deleted	Correlation matrix	No. 15 Reentered	
16.	14.	17.	
-1.24	-.3678945685	-1.24	
1.5376	24.	1.5376	
0.	.1338051988	0.	
7.74	34.	7.74	
	-0.438443332		
	13.		
Parameters	.8327083373	Parameters	
381.5680073	23.	429.2669487	
	-.8442770068		
	12.		
	-.9497122512		
.6875660416		.6567737196	
.1858891041		.1794842476	
-0.119344961		-0.087837239	
.1858891041		.1794842476	
.0727255622		.0713933409	
.0649981587		.0715518196	
-0.119344961		-0.087837239	
.0649981587		.0715518196	
.611395726		.5791559835	
0.913203052	R	.8809931548	R
.3069826983	S	.3571467185	S
6.018571504		6.031048047	
-1.447007221		-1.574435553	
-0.563735175		-.5902404895	
-.9669141557		-.8365252728	
16.	N	17.00	N
0.913203052	R	0.88	R
.3069826983	S	0.36	S
20.0876522	F	15.02	F
6.018571504	A0	6.03	A0
.5645637553	CI	0.63	CI
-1.447007221	A1	-1.57	A1
.5548708989	CI	0.63	CI
-0.563735175	A2	-0.59	A2
.1804588902	CI	0.21	CI
-.9669141557	A3	-0.84	A3
.5232339772	CI	0.59	CI
	R		R
.8312911961	12	0.83	12
-.1840711298	13	-0.14	13
.3082451919	23	0.35	23

time the decimal point is fixed at two places as in the data of the original authors.

Program 435 also provides the complete correlation matrix. Table 4-17 (bottom of column 4) has the results of all 17 points and Table 4-18 (center column) has the results for 16 points.

D. Quintivariate Data: $y = a_0 + a_1x_1 + a_2x_2 + a_3x_3 + a_4x_4$

For this example the data of Table 4-15 is used again, this time with the addition of the F value, column 5 as x_4, the fourth independent variable. Programs 441 and 442 are entered and label A is pressed. The data are entered as before (Table 4-19). Then label B is pressed to begin the calculation. Program 443 is entered and label A is pressed to continue the calculation. An unlabeled printout is obtained (Table 4-19, column 3) with the determinant and matrix components being printed as well as a_1, a_2, a_3, and a_4 (unlabeled). Program 444 is now entered and, when label A is pressed, the statistical data are printed and labeled, as illustrated in the top part of column 4, Table 4-19. Program 445 is then entered and label A is pressed to provide the coefficients of the regression equation (center of column 4). Program 446 is now entered and label A is pressed to obtain the correlation matrix, column 4 of Table 4-19. Note that R_{15}, R_{25}, and R_{35} are the same as R_{14}, R_{24}, and R_{34} of Table 4-16. This is as it should be since the same sets of data are used. Correlation of the new data set (F) with y (R_{45}) is low (0.28), but there is an improvement in R, S, and F with the addition of the variable F. Program 447 can now be entered and used to calculate residuals as is the case with four variables. Program 447 can also be used to delete data. The comparable equation derived by Leclerc and his associates is

$$pD_2^* = -2.08(\pm 0.45)E_{S_{2+6}} - 0.82(\pm 0.17)E_{S_{2+6}}^2$$

$$- 1.28(\pm 0.42)E_{S_2} - 0.48(\pm 0.22F) + 6.11$$

$$n = 17, \quad r = 0.96, \quad S = 0.22, \quad F = 35.8$$

Our value of F is slightly different (35.7), but the other values are all the same. The original authors added yet another term to the regression but the six variable equations had poorer values for r, S, and F.

Table 4-20 provides the residuals (Δ values) for each data point of Table 4-15. The residuals are the differences between the experimental values for pD_2^* and those calculated from the regression equation. The largest residual (-0.37) is for point number 10 (second from the bottom in column 2 of Table 4-20).

TABLE 4-19

Statistical Analysis of Quintivariate Data (Table 4-15)

Data input			Parameters	
1.	8.	15.	6.110786551	
-3.42	-1.24	-1.24	17.	N
11.6964	1.5376	1.5376	0.960425677	R
-1.71	0.	0.	.2188416984	S
-0.18	0.33	-0.09	35.6685047	F
5.79	7.11	7.74		
2.	9.	16.	6.110786551	A0
0.	-2.12	-2.3	.2894312834	CI
0.	4.4944	5.29	-2.07719665	A1
0.	-1.06	-1.06	.4500505554	CI
0.	1.87	0.68	-.8232995392	A2
5.84	7.13	7.74	.1662960597	CI
			-1.28253518	A3
3.	10.	17.	.4166220666	CI
0.	-1.5	-2.48	-.4822621484	A4
0.	2.25	6.1504	.2210140407	CI
0.	-0.75	-1.24		
0.41	1.52	-0.12		
6.09	7.23	7.74	6.11	A0
			0.29	CI
			-2.08	A1
4.	11.		0.45	CI
-2.4	-2.48	1999.81377	-0.82	A2
5.76	6.1504		0.17	CI
0.	-1.24		-1.28	A3
0.35	1.56	Matrix components	0.42	CI
6.13	7.28	.8900630488	-0.48	A4
		.2876274386	0.22	CI
5.	12.	.1191186146		
-0.25	-2.12	.2237774835	Correlation matrix	
0.0625	4.4944	.2876274386	15.	
0.	-1.06	.1215240083	-.3184814951	
0.41	1.75	0.167487911	25.	
6.58	7.37	.1037338966	.0760376813	
		.1191186146	35.	
6.	13.	0.167487911	-.3346782489	
-1.06	-2.62	0.762750859	45.	
1.1236	6.8644	.1985176959	.2751465793	
0.	-1.31	.2237774835	14.	
1.13	-0.14	.1037338966	-.0817570761	
6.71	7.61	.1985176959	24.	
		.2146534618	-.0836650847	
7.	14.		34.	
-1.06	-2.12	-2.07719665	-.1873331615	
1.1236	4.4944	-.8232995392	13.	
0.	-1.06	-1.28253518	.8289543953	
0.71	1.48	-.4822621484	23.	
7.08	7.66		-.8486609946	
			12.	
			-.9487185451	

TABLE 4-20

Residuals for Quintivariate Equation (Table 4-19)

-3.42		-1.06		-2.12	
11.70		1.12		4.49	
-1.71		0.00		-1.06	
-0.18		1.13		1.75	
5.87	YC	6.84	YC	7.33	YC
5.79	Y	6.71	Y	7.37	Y
-0.08	△	-0.13	△	0.04	△
0.00		-1.06		-2.62	
0.00		1.12		6.86	
0.00		0.00		-1.31	
0.00		0.71		-0.14	
6.11	YC	7.05	YC	7.65	YC
5.84	Y	7.08	Y	7.61	Y
-0.27	△	0.03	△	-0.04	△
0.00		-1.24		-2.12	
0.00		1.54		4.49	
0.00		0.00		-1.06	
0.41		0.33		1.48	
5.91	YC	7.26	YC	7.46	YC
6.09	Y	7.11	Y	7.66	Y
0.18	△	-0.15	△	0.20	△
-2.40		-2.12		-1.24	
5.76		4.49		1.54	
0.00		-1.06		0.00	
0.35		1.87		-0.09	
6.19	YC	7.27	YC	7.46	YC
6.13	Y	7.13	Y	7.74	Y
-0.06	△	-0.14	△	0.28	△
-0.25		-1.50		-2.30	
0.06		2.25		5.29	
0.00		-0.75		-1.06	
0.41		1.52		0.68	
6.38	YC	7.60	YC	7.56	YC
6.58	Y	7.23	Y	7.74	Y
0.20	△	-0.37	△	0.18	△
		-2.48		-2.48	
		6.15		6.15	
		-1.24		-1.24	
		1.56		-0.12	
		7.04	YC	7.85	YC
		7.28	Y	7.74	Y
		0.24	△	-0.11	△

Table 4-21 illustrates the removal and the subsequent addition of data point number 10. Removal is shown in column 1. Now there are 16 data points (N) and the correlation coefficient (R), the standard deviation (S), and the F values are all improved over the values in Table 4-19. The 95% confidence intervals (CI) have also improved. Column 2 shows the effect

TABLE 4-21

Data Deletion and Addition for Quintivariate Equation (Table 4-19)

No. 10 Deleted	Correlation matrix	No. 10 Reentered	
16.	15.	17.	
-1.5	-.3242137567	-1.5	
2.25	25.	2.25	
-0.75	0.08781811	-0.75	
1.52	35.	1.52	
7.23	-.3315807031	7.23	
	45.		
	.2609573985		
	14.		
1722.217547	-0.099637491	1999.81377	
	24.		
	-.0510301862		
	34.		
.9014824672	-.1794714117	.8900630488	
0.295188715	13.	.2876274386	
.1380715269	.8334738791	.1191186146	
.2190737077	23.	.2237774835	
0.295188715	-.8625318713	.2876274386	
.1265306468	12.	.1215240083	
.1800374299	-.9510837634	0.167487911	
.1006193292		.1037338966	
.1380715269		.1191186146	
.1800374299		0.167487911	
.7942071794		0.762750859	
.1907107966	Residuals	.1985176959	
.2190737077	-1.24	.2237774835	
.1006193292	1.5376	.1037338966	
.1907107966	0.	.1985176959	
.2165909954	-0.09	.2146534618	
	7.476102357 YC		
	7.74 Y		
-2.121472807	.2638976426 $\hat{}$	-2.07719665	
-.8526166412		-.8232995392	
-1.356020718	-2.3	-1.28253518	
-.4640243446	5.29	-.4822621484	
Parameters	-1.06	Parameters	
6.114697233	0.68	6.110786551	
16.	N	7.605588065 YC	17.	N
.9721826765	R	7.74 Y	0.960425677	R
0.191413332	S	.1344119355 $\hat{}$	.2188416984	S
47.37682685	F		35.6685047	F

6.114697233	A0		6.110786551	A0
.2557712831	C I		.2894312834	C I
-2.121472807	A1		-2.07719665	A1
.4002057521	C I		.4500505554	C I
-.8526166412	A2		-.8232995392	A2
.1499348357	C I		.1662960597	C I
-1.356020718	A3		-1.28253518	A3
.3756397894	C I		.4166220666	C I
-.4640243446	A4		-.4822621484	A4
.1961665177	C I		.2210140407	C I

on the correlation matrix and on two of the residuals. Column 3 shows that the point number 10 may again be added to the data base, the calculations repeated, and the original results of Table 4-19 obtained. Thus the programs provide the investigator with a means of examining data in a variety of ways.

E.   Coefficients of a Polynomial: Generation of t Values

The programs that provide 95% confidence intervals use a Student's t value, which is normally found in tables in statistical texts. By trial-and-error experimentation, it was found that the t values can be generated by a simple polynomial: $t = a_0 + a_1x + a_2x^2 + a_3x^3$, in which the value of x is obtained from the expression $x = [1 - (1/\varphi)]^{-1}$, where φ is the number of degrees of freedom [this is equivalent to $\varphi/(\varphi - 1)$]. The number of degrees of freedom for a regression equation depends on the number of variables and the number of items n in the calculation. It is impractical to determine φ before each calculation, and it is a convenience to have the program determine both φ and t each time the calculation label is pressed. The coefficients of the polynomial can be found using the least squares method for

$$y = a_0 + a_1x_1 + a_2x_2 + a_3x_3$$

in which $x_1 = x$, $x_2 = x^2$, and $x_3 = x^3$. As Davies pointed out [5, p. 270], there is a very high correlation between the variables in a polynomial so that standard errors and confidence intervals will have little significance.

Nine points are selected from the critical values of t provided by Harshbarger [9] (Table 4-22). These cover the range of $\varphi = 2$ to $\varphi = 28$. A value of $\varphi = 1$ cannot be used in our expression for x and higher values will seldom be required for most sets of experimental data, although, as we shall see, our coefficients will fit these well in any case. The corresponding values of x are calculated as well as x^2 and x^3 and entered into programs 431 and 432 as described in the instructions for entering x_1, x_2, and x_3. The y entries are the t values of the Harshbarger table with an alpha level of 0.05 for a nondirectional test [9]. Label B is pressed and then programs 433 and 434 are entered and label A pressed. Table 4-23 provides the printout of the data input in the first column. Column 2 is the printout of the results. The error term arises because in the calculation of S the square root of a negative number is taken. We can thus ignore the standard error and, as a matter of fact, we are fortunate to have obtained useful coefficients in data that are so highly correlated. When there is perfect correlation, these calculator programs fail because the determinate used in the calculations is zero. In this example the determinate is

TABLE 4-22

Correlation of Student's t Values[a] and a
Function of Degrees of Freedom φ

No.	φ	$\dfrac{1}{1 - (1/\varphi)}$	t
1	2	2	4.303
2	3	1.5	3.182
3	4	1.333333333	2.7776
4	5	1.25	2.571
5	8	1.142857143	2.306
6	12	1.090909091	2.179
7	17	1.0625	2.11
8	23	1.045454545	2.069
9	28	1.037037037	2.048

[a] Data from ref. [9].

0.000073 and the calculation succeeds. When the same calculation was attempted using five variables, i.e.,

$$t = a_0 + a_1x + a_2x^2 + a_3x^3 + a_4x^4$$

the calculation failed because the determinate was 5×10^{-15}!

Columns 2 and 3 of Table 23 provide the residuals for the calculation of these selected t values. In every instance the calculated t value is within ± 0.001 of the tabulated value. The coefficients derived in this manner are used in the programs of this book.

IV. INSTRUCTIONS

A. Two Variables—Program 41: $y = a_0 + a_1x_1$

Step 1   Partition: 6 Op 17 (479.59). *Enter* programs 411 and 412.

Step 2   *Enter data:* Press label A. Enter x, press R/S; enter y, press R/S. Data entries are printed and paper advances, ready for entry of next data set. It is not necessary to press A again unless A′, G, or C has been pressed.

Step 3   *Delete data:* Press label A′. Enter x and y as in step 2. Data entries are again printed and the item number indicates deletion.

Step 4   *Calculate:* Press label B. The statistical parameters n, r, S, and F are printed and labeled, then the coefficients a_0 and a_1 each followed by its respective 95% confidence interval. Label C may be pressed again with fixed decimal format if desired.

Step 5 *Calculate y_c and its residual:* Press label C. Then enter x, press R/S and wait for y_c to be calculated and printed. Enter y and press R/S. The residual Δ will be printed and labeled.

Step 6 Data may be added or deleted at any time and label B or C pressed, following any such addition or deletion.

TABLE 4-23

Derivation of Polynomial Equation for Calculating t Values

Data input[a]				Residuals			
1.	9.			2.	X1	1.090909091	X1
2.	1.037037037			4.	X2	1.190082645	X2
4.	1.075445816			8.	X3	1.298271976	X3
8.	1.115277143			4.3030	YC	2.1794	YC
4.303	2.048			4.3030	Y	2.1790	Y
				0.0000	ᴧ	-0.0004	ᴧ
2.				1.5	X1	1.0625	X1
1.5	.0000729749			2.25	X2	1.12890625	X2
2.25				3.375	X3	1.199462891	X3
3.375				3.1822	YC	2.1100	YC
3.182				3.1820	Y	2.1100	Y
	28049.63658			-0.0002	ᴧ	0.0000	ᴧ
	-19520.1688						
3.	4378.572748						
1.333333333	-19520.1688			1.333333333	X1	1.045454545	X1
1.777777778	13603.53682			1.777777778	X2	1.092975207	X2
2.37037037	-3055.245919			2.37037037	X3	1.142655898	X3
2.776	4378.57275			2.7757	YC	2.0685	YC
	-3055.245919			2.7776	Y	2.0690	Y
	686.9757275			0.0019	ᴧ	0.0005	ᴧ
4.							
1.25	.9999964771	?	R	1.25	X1	1.037037037	X1
1.5625	.0024783751	?	S	1.5625	X2	1.075445816	X2
1.953125				1.953125	X3	1.115277143	X3
2.571	.0086280756	?		2.5705	YC	2.0480	YC
5.	1.21336507	?		2.5710	Y	2.0480	Y
1.142857143	1.00559165	?		0.0005	ᴧ	0.0000	ᴧ
1.306122449	-0.269340616	?					
1.49271137				1.142857143	X1	1.016949153	X1
2.306	9.		N	1.306122449	X2	1.034185579	X2
	.9999964771		R	1.49271137	X3	1.051714148	X3
6.	.0024783751		S	2.3067	YC	1.9993	YC
1.090909091	236546.9154		F	2.3060	Y	2.0000	Y
1.190082645				-0.0007	ᴧ	0.0007	ᴧ
1.298271976	Coefficients						
2.179	.0086280756		A0				
	.7407272683		CI				
7.	1.21336507		A1				
1.0625	1.067254746		CI				
1.12890625	1.00559165		A2				
1.199462891	0.743242257		CI				
2.11	-0.269340616		A3				
	.1670225737		CI				
8.			R				
1.045454545	-0.999295729		12				
1.092975207	.9974663513		13				
1.142655898	-.9994235023		23				
2.069							

[a] Table 4-22

B. Three Variables—Program 42

1. Three-Variable Equation: $y = a_0 + a_1x_1 + a_2x_2$

Step 1 *Partition:* 4 Op17 (639.39). *Enter programs* 421, 422, and 423.

Step 2 *Enter Data:* Press label A. Enter x_1, press R/S. Enter x_2, press R/S. Enter y, press R/S. Repeat for each data set. It is not necessary to press A again unless A′, B, or C have been pressed.

Step 3 *Delete data:* Press label A′. Enter each data set as for data entry.

Step 4 *Calculate:* Press label B. The statistical parameters, regression coefficients and their respective 95% confidence limits are printed and labeled.

Step 5 *Calculate y_c and Δ:* Press label C. Enter x_1 and x_2 with R/S after each entry. Wait for y_c to be calculated and printed. Enter y, press R/S. The residual Δ will be printed and labeled.

Step 6. *Correlation matrix:* Enter program 424. Press label A. The correlation matrix R_{13}, R_{23}, R_{12} will be printed and labeled.

2. Two-Variable Equation from Trivariate Data

Step 1 From trivariate Data: *Enter programs* 425 and 426.

Step 2 Equation in x_1 and y: Press label A and label C.

Step 3 Equation in x_2 and y: Press label B and label C.

Step 4 Equation in x_1 and x_2: Press label A and label B. When the second label is pressed, the statistical parameters for the two-variable equations will be calculated and printed. The F factor by which the three-variable equation is an improvement of the two-variable equation will be labeled F-3V.

Step 5 *Calculate y_c and Δ* for the two-variable equation: Press label D. Enter appropriate single variable, wait for y_c or x_{2c}. Enter corresponding y or x_2 and Δ will be calculated and printed.

Step 6 *Further Use of Two-Variable Equation:* Change partition to 479.59 (6 Op17) and enter programs 411 and 412. Two-variable data may now be added and/or deleted as described for two-variable equation. There is no program to change from two-variable to a three-variable equation.

3. Bilinear Equation: $y = a_0 + a_0 \log P + a_2 \log(\beta P + 1)$

Step 1 Partition: 7 Op17 (399.69). *Enter programs* 427 and 428. Initialize: Press label E′.

Step 2 *Enter data: x* and *y*, R/S after each entry. Maximum 14 data sets.

Step 3 *Enter estimated log* βA: If none is known, calculation begins with $\beta = 1$ (log $\beta = 0$), otherwise enter estimated log β and store in R_{69}. Press label A. Calculation proceeds by iteration with each log β, R^2 and F printed. The calculation stops automatically when the maximum F is reached for a change of 0.01 in log β.

Step 4 *Calculate statistical parameters:* Enter programs 421, 422, and 423. Press GTO 309, change 3 to 4. GTO 393 change 3 to 4. GTO 396 change 2 to 3. Press label B to obtain statistical parameters, regression coefficients and 95% confidence limits.

C. Four Variables—Program 43: $y = a_0 + a_1x_1 + a_2x_2 + a_3x_3$

Partition: 6 Op17 (479.59). This program has three parts:

(1)   431, 432—Add data, calculate R and S
(2)   433, 434—Calculate parameters, residuals
(3)   435—Correlation matrix, deletion

1. Programs 431 and 432

Step 1 *Initialize:* Press A.

Step 2 *Enter data* for first point—press R/S after each item: x_1, x_2, x_3, y.

Step 3 *Enter data* for second point—procedure as for Step 2. All other points are entered in the same manner. A zero must be entered for a missing item. When x_1 is entered and R/S is pressed, the x_1^2 appears in the display. If $x_2 = x_1^2$, it is not necessary to reenter x_2. Simply press R/S and then enter x_3 and y in the usual manner.

Step 4 *Calculate:* Press B. a_0, a_1, a_2, and a_3 appear unlabeled. R and S are printed and labeled.

2. Programs 433 and 434

Step 1 *Complete calculation:* Press A. The parameters n, R, S, and F are printed and labeled. Then a_0, a_1, a_2, a_3 are printed and labeled, each followed by its respective 95% confidence interval (CI). Finally, the partial correlation coefficients are printed and labeled R_{12}, R_{13}, R_{23}.

Step 2 *Calculate residuals:* Press B. Enter x_1, x_2, x_3 as requested, R/S x.

3. Program 435

Step 1 *Correlation matrix:* Press A. The correlation matrix will be printed and labeled.

Step 2 *Data deletion:* Press A'. Enter x_1, x_2, and x_3, R/S after each entry. After x_3, reenter programs 431 and 432 and repeat steps for the calculation of parameters.

D.   Five Variables—Program 44: $y = a_0 + a_1x_1 + a_2x_2 + a_3x_3 + a_4x_4$

Partition: 8 Op17 (319.79). This program has six parts:

(1)   441, 442—Add data, begin calculation.
(2)   443—Complete calculation
(3)   444—Parameters
(4)   445—Coefficients
(5)   446—Correlation matrix
(6)   447—Delete data, y_C and Δ.

1. Programs 441 and 442

Step 1 *Initialize:* Press A.

Step 2 *Enter data* as for four variables: x_1, x_2, x_3, x_4, y.

Step 3 *Begin calculation:* Press B. (There is no printout.)

2. Program 443

Step 1 *Complete calculation:* Press A. a_1, c_2, a_3, and a_4 are printed without labels.

3. Program 444

Step 1 *Statistical parameters:* Press A. a_0 is printed without label. This is followed by n, R, S, F, each labeled.

4. Program 445

Step 1 *Coefficients:* Press A. a_0, a_1, a_2, a_3, and a_4 are printed and labeled, each followed by its respective 95% confidence interval.

5. Program 446

Step 1 *Correlation matrix:* Press A. R_{15}, R_{25}, R_{35}, R_{45}, R_{14}, R_{24}, R_{34}, R_{13}, R_{23}, R_{12} are printed and labeled ($5 = y$).

6. Program 447

Step 1 *Delete data:* Press A, then enter each item as in Section IV.C.1., Step 2. After data deletion, the sequence of steps 1–4 must be repeated to obtain the new parameters and coefficients.

Step 2 *Obtain y_C:* Press B. Enter x_1, x_2, x_3, and x_4 with R/S after each item. If x_2 is x_1^2, press GTO 163. Replace R/S with x^2.

Step 3 *Obtain Δ:* Enter y, press R/S. Δ is printed and labeled.

E. Programs for Regression Analysis—Label Operations Summary

Use Master Library Module throughout. Press R/S after each item of data entry.

1. Two Variables: $y = a_0 + a_1x_1$; 6 Op17 (479.59)

411, 412

A—Add data: X, y.
A'—Delete data: X, y.
B—Calculate: n, R, s, F, a_0, CI, a_1, CI.
C—Enter x, obtain y_C; enter y, obtain Δ.

2. Three Variables: $y = a_0 + a_1x_1 + a_2x_2$; 4 Op17 (639.39)

421, 422, 423

A—Add data: x_1, x_2, y.
A'—Delete data: x_1, x_2, y.
B—Calculate: n, R, s, F, a_0, CI, a_1, CI, a_2, CI.
C—Enter X_1, X_2, obtain y_C; enter y, obtain Δ.

Correlation matrix:
    424: A—R_{13}, R_{23}, R_{12}
    $3V \rightarrow 2V$: $y = a_0 + a_1x_1$; $y = a_0 + a_1x_2$; $X_2 = a_0 + a_1X_1$
    425, 426:

A (or B)—Select x_1 (or x_2) as independent variable.
B (or C)—Select x_2 (or y) as dependent variable. Calculate n, r, s, F, F-3V, a_0, CI, a_1, CI.
D—Enter x_1 (or x_2), obtain y_C (or x_{2C}). Enter x_2 (or y), obtain Δ.

Bilinear equation: $y = a_0 + a_1 \log x + a_2(\beta x + 1)$
    427, 428: 7 Op17 (399.69)

E'—Initialize. Enter data: x, y.
A—Calculate: $\log \beta$, R^2, F.

3. *Four Variables:* $y = a_0 + a_1x_1 + a_2x_2 + a_3x_3$;
6 Op17 (479.59)

431, 432

A—Add data: x_1, x_2, x_3, y.
B—Begin calculation.

433, 434

A—Calculate: n, R, S, F, a_0, CI, a_1, CI, a_2, CI, a_3, CI, partial R's: 12, 13, 23.
B—Enter x_1, x_2, x_3, obtain y_C; enter y, obtain Δ.

435

A—Obtain R's: 14, 24, 34, 13, 23, 12.
A'—Delete data: x_1, x_2, x_3, y.

4. *Five Variables:* $y = a_0 + a_1x_1 + a_2x_2 + a_3x_3 + a_4x_4$;
8 Op17 (319.79)

441, 442

A—Add data: x_1, x_2, x_3, x_4, y.
B—Begin calculation.

443: A—Continue calculation.
444: A—Complete calculation: n, R, S, F.
445: A—Print coefficients: a_0, CI, a_1, CI, a_2, CI, a_3, CI, a_4, CI.
446: A—Obtain R's: 15, 25, 35, 45, 14, 24, 34, 13, 23, 12

447

A—Delete data: x_1, x_2, x_3, x_4, y.
B—Enter x_1, x_2, x_3, x_4, obtain y_C. Enter y, obtain Δ.

V. DESIGN

As explained in Section II, these programs are designed to calculate the coefficients of the least squares equations. These coefficients are derived from data in the form of sums: sums of squares and sums of products. Individual data items are stored only long enough to calculate these sums and products. The equations are solved by calculating matrix components from the sums and products and arranging the components in matrix form. The inverse matrix is then obtained with the Master Library Module. Components of the matrix and of the inverse matrix are used to calculate the coefficients of the regression equations, their confidence limits, and the statistical parameters R, s, and F.

More specifically, the programs for linear regression analysis are designed to perform the following functions:

The items of each data point x_1, x_2, . . . , x_4, y are entered and stored temporarily. Only the items of the last data point are retained indefinitely.

The sums, sums of squares, and sums of products are computed and stored, i.e., Σx_1, Σx_1^2, $\Sigma x_1 x_2$, $\Sigma x_2 y$, Σy^2, etc.

The matrix components are computed from the sums $C_{11} = \Sigma x^2 - (\Sigma x)^2/n$, $C_{12} = \Sigma x_1 x_2 - \Sigma x_1 \Sigma x_2/n$, etc., and stored in appropriate registers as required by the Master Library Module.

The components of the inverse matrix are computed by the Master Library Module. The module uses a process known as "pivoting," which may scramble the components of the inverse matrix. For this reason, the Master Library Module is used to recall the components of the inverse matrix and these are stored in defined locations.

The regression coefficients a_0, a_1, . . . , a_4, their 95% confidence limits, and the statistical parameters s, R, and F are calculated, printed, and labeled. For multiple regressions, the simple correlation coefficients are calculated as components of the correlation matrix.

Provision is made for addition or deletion of one or more data points at any time.

Calculated values of the dependent variable y_c may be obtained from x_1, . . . , x_4 and the residuals obtained when y is entered.

The programs are also designed so that a number of special operations may be carried out. For instance, the residual sum of squares and the sum of squares due to the regression may be obtained as described in Section II.

Program 42 is designed to allow the entry of three variables and, after calculation of the coefficients of the three-variable equation, the coefficients and statistical parameters of each of the two-variable equations may be calculated. This feature is especially useful for equations of the form $y = a_0 + a_1 x + a_2 x^2$, where it is of interest to know if the x^2 (or x) term is a significant addition to the two-variable equation.

Another program is designed to compute the value of β (or $\log \beta$) in the "bilinear" equation of the form $y = a_0 + a_1 x + a_2 \log(\beta x + 1)$. This equation occurs frequently in correlations involving $\log P$ in structure–activity studies.

The process of finding an optimum value for $\log \beta$ is one of nonlinear regression and involves an iterative procedure. Figure 4-4 illustrates the decisions made by programs 427 and 428 once raw data are entered. In this case, the same raw data base is used again and again so there is no

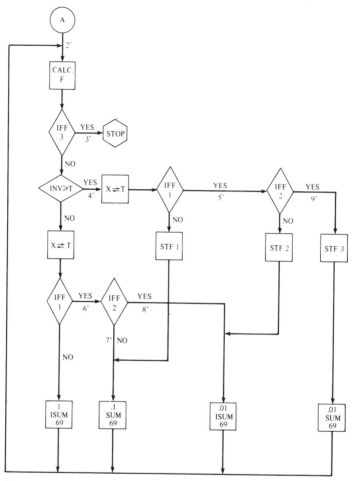

Fig. 4-4 Decision map for program 424, the bilinear equation. When A is pressed, log β is varied in a systematic manner until a minimum F value is obtained.

alternative but to store each item of data. R_{40}–R_{67} are used for this purpose and 14 data points can be accommodated. As illustrated in the example of Table 4-10, R^2 and the F statistic are computed for each value of log β. Log β is varied until F is a maximum. The route taken by the program when label A is pressed may be followed on the decision map (Fig. 4-4). First, a value of F is computed for a given value of log β. This may be selected and stored in R_{69}, or, if none is selected, a value of log $\beta = 0$ is used by the program. None of the flags are set when A is pressed, so, since flag 3 is not set, the program asks if the calculated F

value is less than that stored in the t register (INV $x \geq t$). The t register is made to contain zero when label A is pressed (CP). The F statistic is always a positive number, so in the first instance the answer to INV $x \geq t$ is No. The F value in the display is then placed in the t register ($x \leq t$) to be compared at the next iteration. The program then asks if flag 1 is set. The answer is "no," so log β is changed by subtracting 1 from the initial value in R_{69}.

A new F value is now calculated. Flag 3 is not set so the new F value is compared with the previous value in the t register. If the new F value is equal to or greater than the previous value, iteration continues as before with 1 subtracted from log β on each iteration. Finally, the new F value will be less than the previous one and the answer to INV $x \geq t$ will be "yes." Program operation now goes to label 4'. Direct labels are avoided because they make trouble-shooting difficult in such a complex program.

Flag 1 will not have been set, so it is set by the program and operation goes to label 7'. Here 1 is added to log β in R_{69} and iteration continues. This time the new F value is probably greater than the previous one and flag 1 is set so program operation proceeds to label 6'. The program now asks if flag 2 has been set. It has not, so operation again reaches label 7'. Finally, a value of log β is reached where the new F_2 is less than the previous one, flag 2 is not set so operation sets flag 2.

The iteration proceeds as before except that now 0.01 is subtracted from log β at each iteration. Again a value of the new F is reached, which is less than the previous one. Flag 1 has been set, flag 2 has been set, and flag 3 is finally set. This is the beginning of the last iteration, so 0.01 must now be added to log β to bring F to its maximum value. The new F is again calculated. This time flag 3 has been set, so program operation proceeds to label 3' and the calculation stops.

This program can be quite long in operation but as illustrated in Table 4-10, the printout offers a positive control so the operator can be assured that an optimum log β will finally be reached.

The final statistical parameters are obtained with programs 421, 422, and 423. The number of degrees of freedom ($\varphi = n - 4$) are already computed in R_{20} by programs 427 and 428.

VI. PROGRAMS

A. Register Contents (Table 4-24)

Table 4-24 provides the register contents for all programs for regression analysis. Most of the space is taken up with the sums, sums of

TABLE 4-24
Register Contents
Programs for Regression Analysis

Two variables

	0	1
0	n	R
1	Σy	s
2	Σy^2	$t \to ts$
3	n	x
4	Σx	y
5	Σx^2	a_0
6	Σxy	a_1
7	c_{11}	—
8	c_{1y}	y_c
9	c_{yy}	φ

Three variables

	0	1	2	3
0	n	c_{12}	φ	x_1
1	8	c_{22}	Σx_1	x_2
2	PGM	PGM	Σx_1^2	y
3	PGM	PGM	$\Sigma x_1 x_2$	c_{1y}
4	PGM	c^{11}	Σx_2	c_{2y}
5	PGM	c^{12}	Σx_2^2	c_{yy}
6	PGM	c^{22}	$\Sigma x_1 y$	a_0
7	2	s	$\Sigma x_2 y$	a_1
8	c_{11}	$t \to ts$	Σy	a_2
9	c_{12}	R	Σy^2	y_c

3V → 2V

	0	1	2	3
0	n	R	—	—
1	Σy	—	Σx_1	—
2	Σy^2	$t \to ts$	Σx_1^2	c_{1y}
3	n	x	$\Sigma x_1 x_2$	c_{2y}
4	Σx	y	Σx_2	—
5	Σx^2	a_0	Σx_2^2	—
6	Σxy	a_1	$\Sigma x_1 y$	—
7	c_{11}	s (3V)	$\Sigma x_2 y$	a_1 (3V)
8	c_{1y}	—	Σy	a_2 (3V)
9	c_{yy}	$\varphi \to y_c$	Σy^2	Op 04

Four variables

	0	1	2	3	4	5
0	n	c_{13}	c^{11}	Σx_1	Σx_3^2	c_{3y}
1	8	c_{12}	c^{12}	Σx_1^2	$\Sigma x_3 y$	c_{yy}
2	PGM	c_{22}	c^{13}	$\Sigma x_1 x_2$	Σy	a_0
3	PGM	c_{23}	c^{22}	$\Sigma x_1 x_3$	Σy^2	a_1
4	PGM	c_{13}	c^{23}	$\Sigma x_1 y$	x_1	a_2
5	PGM	c_{23}	c^{33}	Σx_2	x_2	a_3
6	PGM	c_{33}	s	Σx_2^2	x_3	CI a_0
7	3	PGM	$t \to ts$	$\Sigma x_2 x_3$	y	y_c
8	c_{11}	PGM	R	$\Sigma x_2 y$	c_{1y}	Δ
9	c_{12}	PGM	φ	Σx_3	c_{2y}	—

Bilinear equation

	0	1	2	3	4	5	6
0	$0 \to n$	c_{12}	φ	x_1	$x_{(1)}$	:	
1	$Pt \to 8$	c_{22}	Σx_1	x_2	$y_{(1)}$	:	
2	PGM	PGM	Σx_1^2	y	$x_{(2)}$	:	
3	PGM	PGM	$\Sigma x_1 x_2$	c_{1y}	$y_{(2)}$	:	
4	PGM	c^{11}	Σx_2	c_{2y}	:	:	
5	PGM	c^{12}	Σx_2^2	c_{yy}	:	:	
6	PGM	c^{22}	$\Sigma x_1 y$	last F	:	:	$x_{(14)}$
7	2	s	$\Sigma x_2 y$	a_1	:	:	$y_{(14)} \to$ (last F)
8	c_{11}	F	Σy	a_2	:	:	$n \to 0$
9	c_{12}	R	Σy^2	—	:	:	$\log \beta$

Five variables

	0	1	2	3	4	5	6	7
0	n	c_{13}	c_{14}	c^{13}	$\Sigma x_1 x_2$	Σx_4	x_3	a_2
1	8	c_{14}	c_{24}	c^{14}	$\Sigma x_1 x_3$	Σx_4^2	x_4	a_3
2	PGM	c_{12}	c_{34}	c^{22}	$\Sigma x_1 x_4$	Σy	y	a_4
3	PGM	c_{22}	c_{44}	c^{23}	Σx_2^2	Σy^2	c_{1y}	φ
4	PGM	c_{23}	PGM	c^{24}	$\Sigma x_2 x_3$	$\Sigma x_1 y$	c_{2y}	s
5	PGM	c_{24}	PGM	c^{33}	$\Sigma x_2 x_4$	$\Sigma x_2 y$	c_{3y}	R
6	PGM	c_{13}	PGM	c^{34}	$\Sigma x_2 x_4$	$\Sigma x_3 y$	c_{4y}	$t \to ts$
7	4	c_{23}	PGM	c^{44}	Σx_3	$\Sigma x_4 y$	c_{yy}	CI a_0
8	c_{11}	c_{33}	c^{11}	Σx_1	Σx_1^2	x_1	a_0	y_c
9	c_{12}	c_{34}	c^{12}	Σx_1^2	$\Sigma x_3 x_4$	x_2	a_1	Δ

squares and sums of products. Raw data are retained only for the last point entered because new sums are computed for each point as it is added or deleted. The only exception is the program for the bilinear equation for which all raw data must be stored so that new sums can be computed for each iteration. For all other programs, there is no upper limit to the number of data points that may be entered. When calculation begins, matrix components are calculated and stored in proper location for matrix multiplication by the Master Library Module. The numbers in R_{01} and R_{07} are required for matrix multiplication and register contents labeled PGM are left vacant for use of the program of the Master Library Module. The Student's t value is computed when required (see Example E). The coefficients for the regression equation are stored at the end of the memory bank. When calculation is complete, these may be recalled individually, if desired, although when recalled by the program, 95% confidence intervals are calculated as well.

B. Program Listings

Table 4-25 lists all the programs for regression analysis together with the location of the corresponding program listing in Tables 4-26 to 4-47 inclusive, the partitioning required, and the label operations. It should be noted that programs enclosed in parentheses are entered as a unit and are designed to operate that way. For instance, programs 421, 422, and 423 must all be entered for the three-variable equation before label A is pressed to enter raw data. When the data are entered, B is pressed to provide statistical parameters. Now, to obtain the correlation matrix, program 424 is entered and A is pressed. If more data are then to be added or subtracted, program 421 must be reentered first.

To assist in following the programs, those with two and three variables are provided with notes (numbers in parentheses), print symbol designations, and register contents in the margins of the tables. Since the operation of programs for four and five variables is parallel to those of two and three variables, further notes are not required. The general notes of Chapter 1, Section VI apply here also.

C. Program Notes

1. Programs 411, 412

(1)   Label A is used to position the program operation for data entry.

(2)   This program uses the built-in statistics program that automatically stores data in R_{01}–R_{06}.

TABLE 4-25

Regression Analysis Program Summary

Title	No.	Table	Partition (key)	Label operations
Two Variables	411	4-26	477.59 (6 Op 17)⎱	A: Add; A': Delete
	412	4-27	477.59 (6 Op 17)⎰	B: Parameters; C: Residuals
Three Variables	421	4-28	639.39 (4 Op 17)⎱	A: Add; A': Delete
	422	4-29	639.39 (4 Op 17)⎰	B: Parameters
	423	4-30	639.39 (4 Op 17)⎰	
	424	4-31	639.39 (4 Op 17)	A: Correlation matrix
	425	4-32	639.39 (4 Op 17)⎱	A(B): $x_1(x_2)$ Independent var.
	426	4-33	639.39 (4 Op 17)⎰	B(C): $x_2(y)$ Dependent var.
Bilinear	427	4-34	399.69 (7 Op 17)⎱	E': Initialize, add
equation	428	4-35	399.69 (7 Op 17)⎰	A: Parameters
Four Variables	431	4-36	479.59 (6 Op 17)⎱	A: Add
	432	4-37	479.59 (6 Op 17)⎰	B: Begin calculation
	433	4-38	479.59 (6 Op 17)⎱	A: Parameters
	434	4-39	479.59 (6 Op 17)⎰	B: Residuals
	435	4-40	479.59 (6 Op 17)	A: Correlation matrix
				A': Delete
Five Variables	441	4-41	319.79 (8 Op 17)⎱	A: Add
	442	4-42	319.79 (8 Op 17)⎰	B: Begin calculation
	443	4-43	319.79 (8 Op 17)	A: Continue calculation
	444	4-44	319.79 (8 Op 17)	A: n, R, S, F
	445	4-45	319.79 (8 Op 17)	A: Coefficients
	446	4-46	319.79 (8 Op 17)	A: Correlation matrix
	447	4-47	319.79 (8 Op 17)	A: Delete
				B: Residuals

(3)　Label A' positions program operation for data deletion.

(4)　Label B performs the calculations. Some of these steps would be shorter if the Op codes were used.

(5)　The sequence 096–141 calculates t for the φ in R_{19} and 95% confidence interval. See the example of Section III.E. for details of this calculation.

(6)　Label C is used to position program operation to calculate y_c and Δ.

2.　Programs 421, 422, 423

(1)　Label A is used to position program operation for data entry.

(2)　Label A' is used to position program operation for data deletion.

(3)　Use of the matrix program 02 of the Master Library Module requires a 2 in R_{07} and 8 in R_{01} for proper operation. Steps 1 and 2 of that

TABLE 4-26

Program 411 Listing

000	76	LBL	(1)	060	95	=		120	33	X²		180	55	÷	
001	11	A		061	42	STO		121	65	×		181	53	(	
002	91	R/S		062	07	07	C_{11}	122	01	1		182	43	RCL	
003	42	STO		063	43	RCL		123	93	.		183	07	07	C_{11}
004	10	10	x	064	06	06	Σxy	124	00	0		184	65	×	
005	69	OP		065	75	-		125	00	0		185	43	RCL	
006	20	20		066	43	RCL		126	05	5		186	09	09	C_{yy}
007	43	RCL		067	01	01	Σy	127	06	6		187	54	)	
008	00	00	n	068	65	×		128	85	+		188	34	ΓX	
009	99	PRT		069	43	RCL		129	43	RCL		189	95	=	
010	43	RCL		070	04	04	Σx	130	12	12		190	42	STO	
011	10	10	x	071	55	÷		131	45	Yˣ		191	10	10	R
012	99	PRT		072	43	RCL		132	03	3		192	69	OP	
013	32	X:T	(2)	073	03	03	n	133	65	×		193	06	06	print
014	91	R/S		074	95	=		134	93	.		194	03	3	S
015	42	STO		075	42	STO		135	02	2		195	06	6	
016	11	11	y	076	08	08	C_{1y}	136	06	6		196	69	OP	
017	78	Σ+		077	43	RCL		137	09	9		197	04	04	
018	43	RCL		078	02	02	Σy^2	138	94	+/-		198	53	(	
019	11	11		079	75	-		139	95	=		199	53	(	
020	99	PRT		080	43	RCL		140	42	STO		200	43	RCL	
021	98	ADV		081	01	01	Σy	141	12	12	t	201	09	09	C_{yy}
022	61	GTO		082	33	X²		142	43	RCL		202	75	-	
023	11	A	(1)	083	55	÷		143	08	08	C_{1y}	203	43	RCL	
024	76	LBL		084	43	RCL		144	55	÷		204	16	16	a_1
025	16	A'	(3)	085	03	03	n	145	43	RCL		205	65	×	
026	91	R/S		086	95	=		146	07	07	C_{11}	206	43	RCL	
027	42	STO		087	42	STO		147	95	=		207	08	08	C_{1y}
028	10	10	x	088	09	09	C_{yy}	148	42	STO		208	54	)	
029	32	X:T	(2)	089	43	RCL		149	16	16	a_1	209	55	÷	
030	69	OP		090	03	03	n	150	53	(		210	43	RCL	
031	30	30		091	75	-		151	43	RCL		211	19	19	ϕ
032	43	RCL		092	02	2		152	01	01	Σy	212	54	)	
033	00	00		093	95	=		153	75	-		213	34	ΓX	
034	99	PRT		094	42	STO		154	43	RCL		214	49	PRD	tS
035	43	RCL		095	19	19	ϕ	155	16	16	a_1	215	12	12	
036	10	10	x	096	35	1/X	(5)	156	65	×		216	69	OP	
037	99	PRT		097	94	+/-		157	43	RCL		217	06	06	print
038	91	R/S		098	85	+		158	04	04	Σx	218	02	2	F
039	42	STO		099	01	1		159	54	)		219	01	1	
040	11	11	y	100	95	=		160	55	÷		220	69	OP	
041	22	INV		101	35	1/X		161	43	RCL		221	04	04	
042	78	Σ+	delete	102	42	STO		162	03	03	n	222	43	RCL	
043	43	RCL		103	12	12		163	95	=		223	10	10	R
044	11	11	y	104	65	×		164	42	STO		224	33	X²	
045	99	PRT		105	01	1		165	15	15		225	55	÷	
046	98	ADV		106	93	.		166	03	3		226	53	(	
047	61	GTO		107	02	2		167	01	1		227	94	+/-	
048	16	A'	(3)	108	01	1		168	69	OP		228	85	+	
049	76	LBL		109	03	3		169	04	04		229	01	1	
050	12	B	(4)	110	04	4		170	43	RCL		230	54	)	
051	43	RCL		111	85	+		171	03	03		231	65	×	
052	05	05	Σx^2	112	93	.		172	69	OP		232	43	RCL	ϕ
053	75	-		113	00	0		173	06	06	print R	233	19	19	
054	43	RCL		114	00	0		174	03	3		234	95	=	
055	04	04	Σx	115	08	8		175	05	5		235	69	OP	
056	33	X²		116	06	6		176	69	OP		236	06	06	print
057	55	÷		117	85	+		177	04	04		237	98	ADV	
058	43	RCL		118	43	RCL		178	43	RCL		238	01	1	A
059	03	03	n	119	12	12		179	08	08	C_{1y}	239	03	3	

198

TABLE 4-27

Program 412 Listing

240	00	0	O	300	06 06	print
241	01	1		301	98 ADV	
242	69	OP		302	98 ADV	
243	04	04		303	98 ADV	
244	43	RCL		304	91 R/S	
245	15	15	a_0	305	76 LBL	
246	69	OP		306	13 C	(6)
247	06	06	print	307	04 4	x
248	01	1	C	308	04 4	
249	05	5		309	69 OP	
250	02	2	I	310	04 04	
251	04	4		311	91 R/S	
252	69	OP		312	42 STO	
253	04	04		313	13 13	x
254	53	(		314	69 OP	
255	43	RCL		315	06 06	print
256	03	03	n	316	04 4	Y
257	85	+		317	05 5	
258	43	RCL		318	01 1	C
259	04	04	x	319	05 5	
260	33	X²		320	69 OP	
261	55	÷		321	04 04	
262	43	RCL		322	43 RCL	
263	07	07	C_{11}	323	15 15	a_0
264	54	)		324	85 +	
265	34	ΓX		325	43 RCL	
266	55	÷		326	16 16	a_1
267	43	RCL		327	65 ×	
268	03	03	n	328	43 RCL	x
269	65	×		329	13 13	
270	43	RCL		330	95 =	
271	12	12	tS	331	42 STO	
272	95	=		332	19 19	y_c
273	69	OP		333	69 OP	
274	06	06	print	334	06 06	print
275	01	1	A	335	04 4	Y
276	03	3		336	05 5 .	
277	00	0	I	337	69 OP	
278	02	2		338	04 04	
279	69	OP		339	91 R/S	
280	04	04		340	42 STO	
281	43	RCL		341	14 14	y
282	16	16	a_1	342	69 OP	
283	69	OP		343	06 06	print
284	06	06	print	344	07 7	Δ
285	01	1	C	345	05 5	
286	05	5		346	69 OP	
287	02	2	I	347	04 04	
288	04	4		348	43 RCL	
289	69	OP		349	14 14	y
290	04	04		350	75 -	
291	43	RCL	C_{11}	351	43 RCL	
292	07	07		352	19 19	y_c
293	35	1/X		353	95 =	
294	34	ΓX		354	69 OP	
295	65	×		355	06 06	print
296	43	RCL		356	98 ADV	
297	12	12	tS	357	61 GTO	
298	95	=		358	13 C	(6)
299	69	OP		359	00 0	
				360	00 0	

TABLE 4-28
Program 421 Listing

000	76	LBL		060	98	ADV		120	22	INV	
001	11	A (1)		061	61	GTO		121	44	SUM	$\Sigma x_1 y$
002	91	R/S		062	11	A (1)		122	26	26	
003	42	STO		063	76	LBL		123	43	RCL	y
004	30	30 x_1		064	16	A' (2)		124	32	32	
005	69	OP		065	91	R/S		125	65	×	
006	20	20		066	42	STO		126	43	RCL	x_2
007	43	RCL		067	30	30 x_1		127	31	31	
008	00	00 n		068	69	OP		128	95	=	
009	99	PRT		069	30	30		129	22	INV	
010	43	RCL		070	43	RCL		130	44	SUM	$\Sigma x_2 y$
011	30	30 x_1		071	00	00		131	27	27	
012	99	PRT		072	99	PRT		132	98	ADV	
013	44	SUM		073	43	RCL		133	61	GTO	
014	21	21 Σx_1		074	30	30 x_1		134	16	A' (2)	
015	33	X²		075	99	PRT		135	76	LBL	
016	44	SUM		076	22	INV		136	12	B	
017	22	22 Σx_1^2		077	44	SUM		137	02	2	
018	91	R/S		078	21	21 Σx_1		138	42	STO	
019	99	PRT		079	33	X²		139	07	07 (3)	
020	42	STO		080	22	INV		140	08	8	
021	31	31 x_2		081	44	SUM		141	42	STO	
022	44	SUM		082	22	22 Σx_1^2		142	01	01	
023	24	24 Σx_2		083	91	R/S		143	43	RCL	
024	33	X²		084	99	PRT		144	22	22 Σx_1^2	
025	44	SUM		085	42	STO		145	75	-	
026	25	25 Σx_2^2		086	31	31 x_2		146	43	RCL	
027	43	RCL		087	22	INV		147	21	21 Σx_1	
028	30	30 x_1		088	44	SUM		148	33	X²	
029	65	×		089	24	24 Σx_2		149	55	÷	
030	43	RCL		090	33	X²		150	43	RCL	
031	31	31 x_2		091	22	INV		151	00	00 n	
032	95	=		092	44	SUM		152	95	=	
033	44	SUM	$\Sigma x_1 x_2$	093	25	25 Σx_2^2		153	42	STO	
034	23	23		094	43	RCL		154	08	08 C_{11}	
035	91	R/S		095	30	30 x_1		155	43	RCL	
036	99	PRT		096	65	×		156	23	23 $\Sigma x_1 x_2$	
037	42	STO		097	43	RCL		157	75	-	
038	32	32 y		098	31	31 x_2		158	43	RCL	
039	44	SUM	Σy	099	95	=		159	21	21 Σx_1	
040	28	28		100	22	INV		160	65	×	
041	33	X²		101	44	SUM	$\Sigma x_1 x_2$	161	43	RCL	
042	44	SUM		102	23	23		162	24	24 Σx_2	
043	29	29 Σy^2		103	91	R/S		163	55	÷	
044	43	RCL		104	99	PRT		164	43	RCL	
045	32	32 y		105	42	STO		165	00	00 n	
046	65	×		106	32	32 y		166	95	=	
047	43	RCL		107	22	INV		167	42	STO	
048	30	30 x_1		108	44	SUM		168	09	09 C_{12}	
049	95	=		109	28	28 Σy		169	42	STO	
050	44	SUM	$\Sigma x_1 y$	110	33	X²		170	10	10 C_{12}	
051	26	26		111	22	INV		171	43	RCL	
052	43	RCL		112	44	SUM		172	25	25 Σx_2^2	
053	32	32 y		113	29	29 Σy^2		173	75	-	
054	65	×		114	43	RCL		174	43	RCL	
055	43	RCL		115	32	32 y		175	24	24 Σx_2	
056	31	31 x_2		116	65	×		176	33	X²	
057	95	=		117	43	RCL		177	55	÷	
058	44	SUM		118	30	30 x_1		178	43	RCL	
059	27	27 $\Sigma x_2 y$		119	95	=		179	00	00 n	

180	95	=	
181	42	STO	
182	11	11	C_{22}
183	43	RCL	
184	26	26	$\Sigma x_1 y$
185	75	-	
186	43	RCL	
187	21	21	Σx_1
188	65	×	
189	43	RCL	
190	28	28	Σy
191	55	÷	
192	43	RCL	
193	00	00	n
194	95	=	
195	42	STO	
196	33	33	C_{1y}
197	43	RCL	
198	27	27	$\Sigma x_2 y$
199	75	-	
200	43	RCL	
201	24	24	Σx_2
202	65	×	
203	43	RCL	
204	28	28	y
205	55	÷	
206	43	RCL	
207	00	00	n
208	95	=	
209	42	STO	
210	34	34	C_{2y}
211	43	RCL	Σy^2
212	29	29	
213	75	-	
214	43	RCL	Σy
215	28	28	
216	33	X²	
217	55	÷	
218	43	RCL	
219	00	00	n
220	95	=	
221	42	STO	
222	35	35	C_{yy}
223	36	PGM	(4)
224	02	02	
225	13	C	
226	25	CLR	
227	36	PGM	
228	02	02	
229	17	B'	
230	01	1	(5)
231	36	PGM	
232	02	02	
233	18	C'	
234	36	PGM	
235	02	02	
236	91	R/S	
237	42	STO	
238	14	14	C^{11}
239	36	PGM	

TABLE 4-29
Program 422 Listing

Line			Label	Line			Label	Line			Label	Line			Label
240	02	02		300	55	÷		360	05	5	R	420	00	0	
241	91	R/S		301	43	RCL		361	69	OP		421	08	8	
242	42	STO		302	00	00	n	362	04	04		422	06	6	
243	15	15	C^{12}	303	95	=		363	43	RCL		423	85	+	
244	36	PGM		304	42	STO		364	19	19	R	424	43	RCL	
245	02	02		305	36	36	a_o	365	69	OP		425	18	18	
246	91	R/S		306	43	RCL		366	06	06		426	33	X²	
247	36	PGM		307	00	00	n	367	03	3	S	427	65	×	
248	02	02		308	75	-		368	06	6		428	01	1	
249	91	R/S		309	03	3	4 (6)	369	69	OP		429	93	.	
250	42	STO		310	95	=		370	04	04		430	00	0	
251	16	16	C^{22}	311	42	STO		371	43	RCL		431	00	0	
252	43	RCL		312	20	20	ϕ	372	17	17	S	432	05	5	
253	14	14	C^{11}	313	35	1/X		373	69	OP		433	06	6	
254	65	×		314	65	×		374	06	06		434	85	+	
255	43	RCL		315	53	(		375	02	2	F	435	43	RCL	
256	33	33	C_{1y}	316	43	RCL		376	01	1		436	18	18	
257	85	+		317	35	35	C_{yy}	377	69	OP		437	45	YX	
258	43	RCL		318	75	-		378	04	04		438	03	3	
259	15	15	C^{12}	319	43	RCL		379	43	RCL		439	65	×	
260	65	×		320	37	37	a_1	380	19	19	R	440	93	.	
261	43	RCL		321	65	×		381	33	X²		441	02	2	
262	34	34	C_{2y}	322	43	RCL		382	55	÷		442	06	6	
263	95	=		323	33	33	C_{1y}	383	53	(		443	09	9	
264	42	STO		324	75	-		384	94	+/-		444	94	+/-	
265	37	37	a_1	325	43	RCL		385	85	+		445	95	=	
266	43	RCL		326	38	38	a_2	386	01	1		446	42	STO	
267	15	15	C^{12}	327	65	×		387	54	)		447	18	18	t
268	65	×		328	43	RCL		388	65	×		448	43	RCL	
269	43	RCL		329	34	34	C_{2y}	389	53	(		449	17	17	S
270	33	33	C_{1y}	330	54	)		390	43	RCL		450	49	PRD	
271	85	+		331	95	=		391	00	00		451	18	18	tS
272	43	RCL		332	34	ΓX		392	75	-		452	01	1	A
273	16	16	C^{22}	333	42	STO		393	03	3	4 (6)	453	03	3	
274	65	×		334	17	17	S	394	54	)		454	00	0	O
275	43	RCL		335	53	(		395	55	÷		455	01	1	
276	34	34	C_{2y}	336	01	1		396	02	2	3 (7)	456	69	OP	
277	95	=		337	75	-		397	95	=		457	04	04	
278	42	STO		338	43	RCL		398	69	OP		458	43	RCL	
279	38	38	a_2	339	17	17	S	399	06	06		459	36	36	a_o
280	43	RCL		340	33	X²		400	98	ADV		460	69	OP	
281	28	28	Σy	341	55	÷		401	01	1	(8)	461	06	06	
282	55	÷		342	43	RCL		402	75	-		462	01	1	C
283	43	RCL		343	35	35	C_{yy}	403	43	RCL		463	05	5	
284	00	00	n	344	65	×		404	20	20		464	02	2	I
285	75	-		345	43	RCL		405	35	1/X		465	04	4	
286	43	RCL		346	20	20	ϕ	406	95	=		466	69	OP	
287	37	37	a_1	347	54	)		407	35	1/X		467	04	04	
288	65	×		348	34	ΓX		408	42	STO		468	43	RCL	
289	43	RCL		349	42	STO		409	18	18		469	18	18	tS
290	21	21	Σx_1	350	19	19	R	410	65	×		470	55	÷	
291	55	÷		351	03	3	N	411	01	1		471	43	RCL	
292	43	RCL		352	01	1		412	93	.		472	00	00	n
293	00	00	n	353	69	OP		413	02	2		473	65	×	
294	75	-		354	04	04		414	01	1		474	53	(	
295	43	RCL		355	43	RCL		415	03	3		475	43	RCL	
296	38	38	a_2	356	00	00	n	416	04	4		476	00	00	n
297	65	×		357	69	OP		417	85	+		477	85	+	
298	43	RCL		358	06	06		418	93	.		478	43	RCL	
299	24	24	Σx_2	359	03	3		419	00	0		479	14	14	C^{11}

TABLE 4-30

Program 423 Listing

Step			Note		Step			Note		Step			Note
480	65	×			540	69	OP			600	32	32	y
481	43	RCL	Σx_1		541	06	06			601	69	OP	
482	21	21			542	01	1	C		602	06	06	
483	33	X²			543	05	5			603	07	7	Δ
484	85	+			544	02	2	I		604	05	5	
485	43	RCL	C^{22}		545	04	4			605	69	OP	
486	16	16			546	69	OP			606	04	04	
487	65	×			547	04	04			607	43	RCL	
488	43	RCL	Σx_2		548	43	RCL	tS		608	32	32	y
489	24	24			549	18	18			609	75	-	
490	33	X²			550	65	×			610	43	RCL	
491	85	+			551	43	RCL	C^{22}		611	39	39	y_c
492	02	2			552	16	16			612	95	=	
493	65	×			553	34	ГX			613	69	OP	
494	43	RCL	C^{12}		554	95	=			614	06	06	
495	15	15			555	69	OP			615	98	ADV	
496	65	×			556	06	06			616	61	GTO	
497	43	RCL	Σx_1		557	98	ADV			617	13	C	(9)
498	21	21			558	91	R/S			618	00	0	
499	65	×			559	76	LBL			619	00	0	
500	43	RCL	Σx_2		560	13	C	(9)		620	00	0	
501	24	24			561	91	R/S			621	00	0	
502	54	)			562	42	STO	x_1		622	00	0	
503	34	ГX			563	30	30						
504	95	=			564	99	PRT						
505	69	OP			565	91	R/S						
506	06	06			566	42	STO	x_2					
507	01	1	A		567	31	31						
508	03	3			568	99	PRT						
509	00	0	1		569	04	4	Y					
510	02	2			570	05	5						
511	69	OP			571	01	1	C					
512	04	04			572	05	5						
513	43	RCL			573	69	OP						
514	37	37	a_1		574	04	04						
515	69	OP			575	43	RCL	a_0					
516	06	06			576	36	36						
517	01	1	C		577	85	+						
518	05	5			578	43	RCL	a_1					
519	02	2	I		579	37	37						
520	04	4			580	65	×						
521	69	OP			581	43	RCL	x_1					
522	04	04			582	30	30						
523	43	RCL			583	85	+						
524	18	18	tS		584	43	RCL	a_2					
525	65	×			585	38	38						
526	43	RCL	C^{11}		586	65	×						
527	14	14			587	43	RCL	x_2					
528	34	ГX			588	31	31						
529	95	=			589	95	=						
530	69	OP			590	42	STO	y_c					
531	06	06			591	39	39						
532	01	1	A		592	69	OP						
533	03	3			593	06	06						
534	00	0	2		594	04	4	Y					
535	03	3			595	05	5						
536	69	OP			596	69	OP						
537	04	04			597	04	04						
538	43	RCL	a_2		598	91	R/S						
539	38	38			599	42	STO						

TABLE 4-31

Program 424 Listing

000	76	LBL		060	02	02		
001	11	A		061	43	RCL		
002	01	1	(1)	062	23	23	$\Sigma x_1 x_2$	
003	03	3		063	42	STO		
004	99	PRT	(2)	064	06	06		
005	43	RCL		065	69	OP		
006	28	28	Σy	066	13	13	(3)	
007	42	STO		067	99	PRT		
008	01	01		068	91	R/S		
009	43	RCL		069	00	0		
010	29	29	y^2	070	00	0		
011	42	STO		071	00	0		
012	02	02		072	00	0		
013	43	RCL		073	00	0		
014	00	00	n	074	00	0		
015	42	STO		075	00	0		
016	03	03		076	00	0		
017	43	RCL		077	00	0		
018	21	21	Σx_1^2	078	00	0		
019	42	STO		079	00	0		
020	04	04						
021	43	RCL						
022	22	22	Σx_1^2					
023	42	STO						
024	05	05						
025	43	RCL						
026	26	26	$\Sigma x_1 y$					
027	42	STO						
028	06	06						
029	69	OP						
030	13	13	(3)					
031	99	PRT						
032	02	2						
033	03	3						
034	99	PRT						
035	43	RCL						
036	24	24	Σx_2					
037	42	STO						
038	04	04						
039	43	RCL						
040	25	25	Σx_2^2					
041	42	STO						
042	05	05						
043	43	RCL						
044	27	27	$\Sigma x_2 y$					
045	42	STO						
046	06	06						
047	69	OP						
048	13	13						
049	99	PRT						
050	01	1	(3)					
051	02	2						
052	99	PRT						
053	43	RCL						
054	21	21	Σx_1					
055	42	STO						
056	01	01						
057	43	RCL						
058	22	22	Σx_1^2					
059	42	STO						

TABLE 4-32
Program 425 Listing

Addr			Note	Addr			Note	Addr			Note	Addr			Note
000	76	LBL		060	30	30		120	09	09	C_{yy}	180	42	STO	
001	11	A	(1)	061	43	RCL		121	43	RCL		181	16	16	a_1
002	04	4	X	062	24	24	Σx_2	122	03	03	n	182	53	(	
003	04	4		063	42	STO		123	75	-		183	43	RCL	
004	00	0		064	01	01		124	02	2		184	01	01	Σy
005	02	2		065	43	RCL		125	95	=		185	75	-	
006	00	0		066	25	25	Σx_2^2	126	42	STO		186	43	RCL	
007	00	0		067	42	STO		127	19	19	ϕ	187	16	16	a_1
008	00	0		068	02	02		128	35	1/X	(6)	188	65	×	
009	00	0		069	43	RCL		129	94	+/-		189	43	RCL	
010	42	STO		070	23	23	$\Sigma x_1 x_2$	130	85	+		190	04	04	Σx
011	30	30	(2)	071	42	STO		131	01	1		191	54	)	
012	43	RCL		072	06	06		132	95	=		192	55	÷	
013	21	21	Σx_1	073	61	GTO		133	35	1/X		193	43	RCL	
014	42	STO		074	36	PGM	(5)	134	42	STO		194	03	03	n
015	04	04		075	76	LBL		135	12	12		195	95	=	
016	43	RCL		076	36	PGM	(5)	136	65	×		196	42	STO	
017	22	22	Σx_1^2	077	43	RCL		137	01	1		197	15	15	a_0
018	42	STO		078	30	30	(2)	138	93	.		198	03	3	
019	05	05		079	69	OP		139	02	2		199	01	1	N
020	43	RCL		080	04	04		140	01	1		200	69	OP	
021	00	00		081	69	OP		141	03	3		201	04	04	
022	42	STO		082	05	05	print	142	04	4		202	43	RCL	
023	03	03		083	43	RCL		143	85	+		203	03	03	n
024	86	STF		084	05	05	Σx^2	144	93	.		204	69	OP	
025	01	01	(3)	085	75	-		145	00	0		205	06	06	
026	91	R/S		086	43	RCL		146	00	0		206	03	3	R
027	76	LBL		087	04	04	Σx	147	08	8		207	05	5	
028	12	B	(4)	088	33	X²		148	06	6		208	69	OP	
029	87	IFF		089	55	÷		149	85	+		209	04	04	
030	01	01		090	43	RCL		150	43	RCL		210	43	RCL	
031	88	DMS		091	03	03	n	151	12	12		211	08	08	C_{1y}
032	04	4	X	092	95	=		152	33	X²		212	55	÷	
033	04	4		093	42	STO		153	65	×		213	53	(	
034	00	0	2	094	07	07	C_{11}	154	01	1		214	43	RCL	
035	03	3		095	43	RCL		155	93	.		215	07	07	C_{11}
036	00	0		096	06	06	xy	156	00	0		216	65	×	
037	00	0		097	75	-		157	00	0		217	43	RCL	
038	42	STO		098	43	RCL		158	05	5.		218	09	09	C_{yy}
039	30	30	(2)	099	01	01	y	159	06	6		219	54	)	
040	43	RCL		100	65	×		160	85	+		220	34	ΓX	
041	24	24	Σx_2^2	101	43	RCL		161	43	RCL		221	95	=	
042	42	STO		102	04	04	x	162	12	12		222	42	STO	
043	04	04		103	55	÷		163	45	Yˣ		223	10	10	R
044	43	RCL		104	43	RCL		164	03	3		224	69	OP	
045	25	25	Σx_2^2	105	03	03	n	165	65	×		225	06	06	
046	42	STO		106	95	=		166	93	.		226	03	3	
047	05	05		107	42	STO		167	02	2		227	06	6	S
048	43	RCL		108	08	08	C_{1y}	168	06	6		228	69	OP	
049	00	00	n	109	43	RCL		169	09	9		229	04	04	
050	42	STO		110	02	02	Σy^2	170	94	+/-		230	53	(	
051	03	03		111	75	-		171	95	=		231	53	(	
052	91	R/S		112	43	RCL		172	42	STO		232	43	RCL	
053	76	LBL		113	01	01	Σy	173	12	12	t	233	09	09	C_{yy}
054	88	DMS		114	33	X²		174	43	RCL		234	75	-	
055	04	4	X	115	55	÷		175	08	08	C_{1y}	235	43	RCL	
056	04	4		116	43	RCL		176	55	÷		236	16	16	a_1
057	00	0	2	117	03	03	n	177	43	RCL		237	65	×	
058	03	3		118	95	=		178	07	07	C_{11}	238	43	RCL	
059	44	SUM		119	42	STO		179	95	=		239	08	08	C_{1y}

TABLE 4-33

Program 426 Listing

240	54	)	
241	55	÷	
242	43	RCL	
243	19	19	ϕ
244	54	)	
245	34	ΓΧ	
246	49	PRD	
247	12	12	tS
248	69	□P	
249	06	06	print
250	02	2	S
251	01	1	F
252	69	□P	
253	04	04	
254	43	RCL	
255	10	10	R
256	33	X²	
257	55	÷	
258	53	(	
259	94	+/-	
260	85	+	
261	01	1	
262	54	)	
263	65	×	
264	43	RCL	
265	19	19	ϕ
266	95	=	
267	69	□P	
268	06	06	print
269	61	GT□	
270	89	π̃	(7)
271	76	LBL	(8)
272	77	GE	
273	98	ADV	
274	01	1	A
275	03	3	O
276	00	0	
277	01	1	
278	69	□P	
279	04	04	
280	43	RCL	
281	15	15	a_0
282	69	□P	
283	06	06	
284	01	1	C
285	05	5	
286	02	2	I
287	04	4	
288	69	□P	
289	04	04	
290	53	(	n
291	43	RCL	
292	03	03	
293	85	+	
294	43	RCL	
295	04	04	x
296	33	X²	
297	55	÷	
298	43	RCL	
299	07	07	C_{11}

300	54	)	
301	34	ΓΧ	
302	55	÷	
303	43	RCL	
304	03	03	n
305	65	×	
306	43	RCL	
307	12	12	tS
308	95	=	
309	69	□P	
310	06	06	
311	01	1	A
312	03	3	
313	00	0	I
314	02	2	
315	69	□P	
316	04	04	
317	43	RCL	
318	16	16	a_1
319	69	□P	
320	06	06	
321	01	1	C
322	05	5	
323	02	2	I
324	04	4	
325	69	□P	
326	04	04	
327	43	RCL	
328	07	07	C_{11}
329	35	1/X	
330	34	ΓΧ	
331	65	×	
332	43	RCL	
333	12	12	tS
334	95	=	
335	69	□P	
336	06	06	
337	22	INV	
338	86	STF	
339	01	01	
340	98	ADV	
341	98	ADV	
342	98	ADV	
343	91	R/S	
344	76	LBL	
345	13	C	(9)
346	86	STF	
347	02	02	(10)
348	04	4	Y
349	05	5	
350	44	SUM	(2)
351	30	30	
352	43	RCL	
353	28	28	Σy
354	42	ST□	
355	01	01	
356	43	RCL	
357	29	29	Σy^2
358	42	ST□	
359	02	02	

360	87	IFF	
361	01	01	(11)
362	87	IFF	
363	43	RCL	
364	27	27	Σy^2
365	42	ST□	
366	06	06	
367	61	GT□	
368	36	PGM	(5)
369	76	LBL	
370	87	IFF	(12)
371	43	RCL	
372	26	26	$\Sigma x_1 y$
373	42	ST□	
374	06	06	
375	61	GT□	
376	36	PGM	(5)
377	76	LBL	
378	89	π̃	(13)
379	02	2	F
380	01	1	
381	02	2	
382	00	0	
383	00	0	3
384	04	4	
385	04	4	V
386	02	2	
387	69	□P	
388	04	04	
389	53	(	
390	43	RCL	
391	37	37	$a_1(3V)$
392	65	×	
393	43	RCL	
394	33	33	$C_{1y}(3V)$
395	85	+	
396	43	RCL	
397	38	38	$a_2(3V)$
398	65	×	
399	43	RCL	
400	34	34	$C_{2y}(3V)$
401	75	-	
402	43	RCL	
403	08	08	$C_{1y}(2V)$
404	33	X²	
405	55	÷	
406	43	RCL	
407	07	07	$C_{11}(2V)$
408	54	)	
409	55	÷	
410	43	RCL	
411	17	17	$S(3V)$
412	33	X²	
413	95	=	
414	69	□P	
415	06	06	
416	61	GT□	
417	77	GE	
418	76	LBL	
419	14	D	(14)

420	04	4	
421	04	4	X
422	69	□P	
423	04	04	
424	91	R/S	
425	42	ST□	
426	13	13	x
427	69	□P	
428	06	06	
429	04	4	Y
430	05	5	
431	01	1	C
432	05	5	
433	69	□P	
434	04	04	
435	43	RCL	
436	15	15	a_0
437	85	+	
438	43	RCL	
439	16	16	a_1
440	65	×	
441	43	RCL	
442	13	13	x
443	95	=	
444	42	ST□	
445	19	19	y_c
446	69	□P	
447	06	06	
448	04	4	Y
449	05	5	
450	69	□P	
451	04	04	
452	91	R/S	
453	42	ST□	
454	14	14	y
455	69	□P	
456	06	06	
457	07	7	
458	05	5	
459	69	□P	
460	04	04	
461	43	RCL	
462	14	14	y
463	75	-	
464	43	RCL	
465	19	19	y_c
466	95	=	
467	69	□P	
468	06	06	
469	98	ADV	
470	61	GT□	
471	14	D	
472	00	0	
473	00	0	
474	00	0	

TABLE 4-34

Program 427 Listing

000	91	R/S		060	22	INV		120	22	22	
001	42	STO		061	28	LOG		121	75	-	
002	30	30	x_1	062	65	×		122	43	RCL	
003	72	ST*	(1)	063	43	RCL		123	21	21	
004	01	01		064	69	69	log	124	33	X²	
005	69	OP		065	22	INV		125	55	÷	
006	20	20	n	066	28	LOG		126	43	RCL	
007	43	RCL		067	85	+		127	00	00	
008	30	30		068	01	1		128	95	=	
009	99	PRT		069	54	)		129	42	STO	
010	69	OP		070	28	LOG		130	08	08	C_{11}
011	21	21		071	42	STO		131	43	RCL	
012	91	R/S		072	31	31	x_2	132	23	23	
013	99	PRT		073	44	SUM		133	75	-	
014	72	ST*	(1)	074	24	24	(6)	134	43	RCL	
015	01	01		075	33	X²		135	21	21	
016	69	OP		076	44	SUM		136	65	×	
017	21	21		077	25	25		137	43	RCL	
018	98	ADV		078	43	RCL		138	24	24	
019	61	GTO		079	30	30		139	55	÷	
020	00	00		080	65	×		140	43	RCL	
021	00	00		081	43	RCL		141	00	00	
022	76	LBL		082	31	31		142	95	=	
023	11	A	(2)	083	95	=		143	42	STO	
024	29	CP		084	44	SUM		144	09	09	C_{12}
025	43	RCL		085	23	23		145	42	STO	
026	00	00	n	086	69	OP		146	10	10	
027	42	STO		087	21	21		147	43	RCL	
028	68	68	(3)	088	73	RC*		148	25	25	
029	04	4		089	01	01		149	75	-	
030	69	OP		090	42	STO		150	43	RCL	
031	17	17		091	32	32	y	151	24	24	
032	47	CMS		092	44	SUM		152	33	X²	
033	07	7		093	28	28		153	55	÷	
034	69	OP		094	33	X²		154	43	RCL	
035	17	17		095	44	SUM		155	00	00	
036	43	RCL		096	29	29		156	95	=	
037	68	68		097	43	RCL		157	42	STO	
038	42	STO		098	32	32		158	11	11	C_{22}
039	00	00	n	099	65	×		159	43	RCL	
040	04	4		100	43	RCL		160	26	26	
041	00	0		101	30	30		161	75	-	
042	42	STO		102	95	=		162	43	RCL	
043	01	01		103	44	SUM		163	21	21	
044	76	LBL		104	26	26		164	65	×	
045	88	DMS	(4)	105	43	RCL		165	43	RCL	
046	73	RC*		106	31	31		166	28	28	
047	01	01		107	65	×		167	55	÷	
048	67	EQ		108	43	RCL		168	43	RCL	
049	87	IFF		109	32	32		169	00	00	
050	42	STO		110	95	=		170	95	=	
051	30	30	x_1	111	44	SUM		171	42	STO	
052	44	SUM		112	27	27		172	33	33	C_{1y}
053	21	21		113	69	OP		173	43	RCL	
054	33	X²		114	21	21		174	27	27	
055	44	SUM		115	61	GTO		175	75	-	
056	22	22		116	88	DMS	(4)	176	43	RCL	
057	53	(	(5)	117	76	LBL		177	24	24	
058	43	RCL		118	87	IFF	(7)	178	65	×	
059	30	30		119	43	RCL		179	43	RCL	

180	28	28	
181	55	÷	
182	43	RCL	
183	00	00	
184	95	=	
185	42	STO	
186	34	34	C_{2y}
187	43	RCL	
188	29	29	
189	75	-	
190	43	RCL	
191	28	28	
192	33	X²	
193	55	÷	
194	43	RCL	
195	00	00	
196	95	=	
197	42	STO	
198	35	35	C_{yy}
199	02	2	
200	42	STO	
201	07	07	
202	08	8	
203	42	STO	
204	01	01	
205	36	PGM	
206	02	02	
207	13	C	
208	25	CLR	
209	36	PGM	
210	02	02	
211	17	B'	
212	01	1	
213	36	PGM	
214	02	02	
215	18	C'	
216	36	PGM	
217	02	02	
218	91	R/S	
219	42	STO	
220	14	14	C^{11}
221	36	PGM	
222	02	02	
223	91	R/S	
224	42	STO	
225	15	15	C^{12}
226	36	PGM	
227	02	02	
228	91	R/S	
229	36	PGM	
230	02	02	
231	91	R/S	
232	42	STO	
233	16	16	C^{22}
234	43	RCL	
235	14	14	
236	65	×	
237	43	RCL	
238	33	33	
239	85	+	

TABLE 4-35

Program 428 Listing

240	43	RCL	300	43	RCL	360	02	02	
241	15	15	301	18	18	361	68	NOP	
242	65	×	302	65	×	362	76	LBL	
243	43	RCL	303	43	RCL	363	67	EQ	
244	34	34	304	20	20	364	93	.	
245	95	=	305	55	÷	365	01	1	
246	42	STO	306	03	3	366	44	SUM	
247	37	37 a_1	307	95	=	367	69	69	
248	43	RCL	308	42	STO	368	61	GTO	
249	15	15	309	18	18 F	369	11	A	
250	65	×	310	29	CP	370	76	LBL	
251	43	RCL	311	99	PRT	371	68	NOP	
252	33	33	312	87	IFF	372	93	.	
253	85	+	313	03	03 (8)	373	00	0	
254	43	RCL	314	89	ʃ	374	01	1	
255	16	16	315	43	RCL	375	22	INV	
256	65	×	316	67	67 last F	376	44	SUM	
257	43	RCL	317	32	X⫶T (9)	377	69	69	
258	34	34	318	43	RCL	378	61	GTO	
259	95	=	319	18	18	379	11	A	
260	42	STO	320	22	INV	380	76	LBL	
261	38	38 a_2	321	77	GE	381	69	OP	
262	43	RCL	322	77	GE	382	86	STF	
263	69	69 log	323	42	STO	383	03	03	
264	99	PRT	324	67	67	384	93	.	
265	43	RCL	325	87	IFF	385	00	0	
266	00.	00	326	01	01	386	01	1	
267	75	-	327	79	x̄	387	44	SUM	
268	04	4 (7)	328	01	1	388	69	69	
269	95	=	329	22	INV	389	61	GTO	
270	42	STO	330	44	SUM	390	11	A	
271	20	20 φ	331	69	69	391	76	LBL	
272	53	(	332	61	GTO	392	10	E'	
273	43	RCL	333	11	A	393	47	CMS	
274	35	35	334	76	LBL	394	04	4	
275	75	-	335	89	ʃ	395	00	0	
276	43	RCL	336	91	R/S	396	42	STO	
277	37	37	337	76	LBL	397	01	01	
278	65	×	338	77	GE	398	29	CP	
279	43	RCL	339	42	STO	399	81	RST	
280	33	33	340	67	67				
281	75	-	341	87	IFF		Labels		
282	43	RCL	342	01	01				
283	38	38	343	78	Σ+	023	11	A	
284	65	×	344	86	STF	045	88	DMS	
285	43	RCL	345	01	01	118	87	IFF	
286	34·	34	346	61	GTO	335	89	ʃ	
287	54	)	347	67	EQ	338	77	GE	
288	55	÷	348	76	LBL	349	78	Σ+	
289	43	RCL	349	78	Σ+	358	79	x̄	
290	35	35	350	87	IFF	363	67	EQ	
291	95	=	351	02	02	371	68	NOP	
292	42	STO	352	69	OP	381	69	OP	
293	18	18	353	86	STF	392	10	E'	
294	94	+/-	354	02	02				
295	85	+	355	61	GTO				
296	01	1	356	68	NOP				
297	95	=	357	76	LBL				
298	99	PRT R	358	79	x̄				
299	55	÷	359	87	IFF				

TABLE 4-36

Program 431 Listing

000	76	LBL	060	91	R/S	120	30	30	180	55	÷
001	11	A	061	99	PRT	121	65	×	181	43	RCL
002	91	R/S	062	42	STO	122	43	RCL	182	00	00
003	42	STO	063	47	47	123	35	35	183	95	=
004	44	44	064	44	SUM	124	55	÷	184	42	STO
005	69	OP	065	42	42	125	43	RCL	185	48	48
006	20	20	066	33	X²	126	00	00	186	43	RCL
007	43	RCL	067	44	SUM	127	95	=	187	43	43
008	00	00	068	43	43	128	42	STO	188	75	-
009	99	PRT	069	43	RCL	129	09	09	189	43	RCL
010	43	RCL	070	47	47	130	42	STO	190	42	42
011	44	44	071	65	×	131	11	11	191	33	X²
012	99	PRT	072	43	RCL	132	43	RCL	192	55	÷
013	44	SUM	073	44	44	133	36	36	193	43	RCL
014	30	30	074	95	=	134	75	-	194	00	00
015	33	X²	075	44	SUM	135	43	RCL	195	95	=
016	44	SUM	076	34	34	136	35	35	196	42	STO
017	31	31	077	43	RCL	137	33	X²	197	51	51
018	91	R/S	078	47	47	138	55	÷	198	43	RCL
019	99	PRT	079	65	×	139	43	RCL	199	37	37
020	42	STO	080	43	RCL	140	00	00	200	75	-
021	45	45	081	45	45	141	95	=	201	43	RCL
022	44	SUM	082	95	=	142	42	STO	202	35	35
023	35	35	083	44	SUM	143	12	12	203	65	×
024	33	X²	084	38	38	144	43	RCL	204	43	RCL
025	44	SUM	085	43	RCL	145	33	33	205	39	39
026	36	36	086	47	47	146	75	-	206	55	÷
027	43	RCL	087	65	×	147	43	RCL	207	43	RCL
028	45	45	088	43	RCL	148	30	30	208	00	00
029	65	×	089	46	46	149	65	×	209	95	=
030	43	RCL	090	95	=	150	43	RCL	210	42	STO
031	44	44	091	44	SUM	151	39	39	211	13	13
032	95	=	092	41	41	152	55	÷	212	42	STO
033	44	SUM	093	98	ADV	153	43	RCL	213	15	15
034	32	32	094	61	GTO	154	00	00	214	43	RCL
035	91	R/S	095	11	A	155	95	=	215	38	38
036	99	PRT	096	76	LBL	156	42	STO	216	75	-
037	42	STO	097	12	B	157	10	10	217	43	RCL
038	46	46	098	03	3	158	42	STO	218	35	35
039	44	SUM	099	42	STO	159	14	14	219	65	×
040	39	39	100	07	07	160	43	RCL	220	43	RCL
041	33	X²	101	08	8	161	40	40	221	42	42
042	44	SUM	102	42	STO	162	75	-	222	55	÷
043	40	40	103	01	01	163	43	RCL	223	43	RCL
044	43	RCL	104	43	RCL	164	39	39	224	00	00
045	46	46	105	31	31	165	33	X²	225	95	=
046	65	×	106	75	-	166	55	÷	226	42	STO
047	43	RCL	107	43	RCL	167	43	RCL	227	49	49
048	44	44	108	30	30	168	00	00	228	43	RCL
049	95	=	109	33	X²	169	95	=	229	41	41
050	44	SUM	110	55	÷	170	42	STO	230	75	-
051	33	33	111	43	RCL	171	16	16	231	43	RCL
052	43	RCL	112	00	00	172	43	RCL	232	39	39
053	46	46	113	95	=	173	34	34	233	65	×
054	65	×	114	42	STO	174	75	-	234	43	RCL
055	43	RCL	115	08	08	175	43	RCL	235	42	42
056	45	45	116	43	RCL	176	30	30	236	55	÷
057	95	=	117	32	32	177	65	×	237	43	RCL
058	44	SUM	118	75	-	178	43	RCL	238	00	00
059	37	37	119	43	RCL	179	42	42	239	95	=

TABLE 4-37

Program 432 Listing

240	42	STD	300	65	×	360	65	×	420	42	STD
241	50	50	301	43	RCL	361	43	RCL	421	26	26
242	36	PGM	302	49	49	362	30	30	422	53	(
243	02	02	303	85	+	363	55	÷	423	01	1
244	13	C	304	43	RCL	364	43	RCL	424	75	-
245	25	CLR	305	22	22	365	00	00	425	43	RCL
246	36	PGM	306	65	×	366	75	-	426	26	26
247	02	02	307	43	RCL	367	43	RCL	427	33	X²
248	17	B'	308	50	50	368	54	54	428	65	×
249	01	1	309	95	=	369	65	×	429	43	RCL
250	36	PGM	310	42	STD	370	43	RCL	430	29	29
251	02	02	311	53	53	371	35	35	431	55	÷
252	18	C'	312	43	RCL	372	55	÷	432	43	RCL
253	36	PGM	313	21	21	373	43	RCL	433	51	51
254	02	02	314	65	×	374	00	00	434	54	)
255	91	R/S	315	43	RCL	375	75	-	435	34	√X
256	42	STD	316	48	48	376	43	RCL	436	42	STD
257	20	20	317	85	+	377	55	55	437	28	28
258	36	PGM	318	43	RCL	378	65	×	438	98	ADV
259	02	02	319	23	23	379	43	RCL	439	69	OP
260	91	R/S	320	65	×	380	39	39	440	00	00
261	42	STD	321	43	RCL	381	55	÷	441	03	3
262	21	21	322	49	49	382	43	RCL	442	05	5
263	36	PGM	323	85	+	383	00	00	443	69	OP
264	02	02	324	43	RCL	384	95	=	444	04	04
265	91	R/S	325	24	24	385	42	STD	445	43	RCL
266	42	STD	326	65	×	386	52	52	446	28	28
267	22	22	327	43	RCL	387	43	RCL	447	69	OP
268	36	PGM	328	50	50	388	00	00	448	06	06
269	02	02	329	95	=	389	75	-	449	03	3
270	91	R/S	330	42	STD	390	04	4	450	06	6
271	36	PGM	331	54	54	391	95	=	451	69	OP
272	02	02	332	43	RCL	392	42	STD	452	04	04
273	91	R/S	333	22	22	393	29	29	453	43	RCL
274	42	STD	334	65	×	394	35	1/X	454	26	26
275	23	23	335	43	RCL	395	65	×	455	69	OP
276	36	PGM	336	48	48	396	53	(	456	06	06
277	02	02	337	85	+	397	43	RCL	457	98	ADV
278	91	R/S	338	43	RCL	398	51	51	458	43	RCL
279	42	STD	339	24	24	399	75	-	459	52	52
280	24	24	340	65	×	400	43	RCL	460	99	PRT
281	36	PGM	341	43	RCL	401	53	53	461	43	RCL
282	02	02	342	49	49	402	65	×	462	53	53
283	91	R/S	343	85	+	403	43	RCL	463	99	PRT
284	36	PGM	344	43	RCL	404	48	48	464	43	RCL
285	02	02	345	25	25	405	75	-	465	54	54
286	91	R/S	346	65	×	406	43	RCL	466	99	PRT
287	36	PGM	347	43	RCL	407	54	54	467	43	RCL
288	02	02	348	50	50	408	65	×	468	55	55
289	91	R/S	349	95	=	409	43	RCL	469	99	PRT
290	42	STD	350	42	STD	410	49	49	470	91	R/S
291	25	25	351	55	55	411	75	-	471	00	0
292	43	RCL	352	43	RCL	412	43	RCL	472	00	0
293	20	20	353	42	42	413	55	55	473	00	0
294	65	×	354	55	÷	414	65	×			
295	43	RCL	355	43	RCL	415	43	RCL			
296	48	48	356	00	00	416	50	50			
297	85	+	357	75	-	417	54	)			
298	43	RCL	358	43	RCL	418	95	=			
299	21	21	359	53	53	419	34	√X			

TABLE 4-38

Program 433 Listing

000	76	LBL	060	35	1/X	120	43	RCL	180	95	=
001	11	A	061	42	STO	121	23	23	181	69	OP
002	69	OP	062	27	27	122	65	×	182	06	06
003	00	00	063	65	×	123	43	RCL	183	01	1
004	03	3	064	01	1	124	35	35	184	03	3
005	01	1	065	93	.	125	33	X²	185	00	0
006	69	OP	066	02	2	126	85	+	186	03	3
007	04	04	067	01	1	127	43	RCL	187	69	OP
008	43	RCL	068	03	3	128	25	25	188	04	04
009	00	00	069	04	4	129	65	×	189	43	RCL
010	69	OP	070	85	+	130	43	RCL	190	54	54
011	06	06	071	93	.	131	39	39	191	69	OP
012	03	3	072	00	0	132	33	X²	192	06	06
013	05	5	073	00	0	133	54	)	193	01	1
014	69	OP	074	08	8	134	34	ГX	194	05	5
015	04	04	075	06	6	135	95	=	195	02	2
016	43	RCL	076	85	+	136	42	STO	196	04	4
017	28	28	077	43	RCL	137	56	56	197	69	OP
018	69	OP	078	27	27	138	01	1	198	04	04
019	06	06	079	33	X²	139	03	3	199	43	RCL
020	03	3	080	65	×	140	00	0	200	27	27
021	06	6	081	01	1	141	01	1	201	65	×
022	69	OP	082	93	.	142	69	OP	202	43	RCL
023	04	04	083	00	0	143	04	04	203	23	23
024	43	RCL	084	00	0	144	43	RCL	204	34	ГX
025	26	26	085	05	5	145	52	52	205	95	=
026	69	OP	086	06	6	146	69	OP	206	69	OP
027	06	06	087	85	+	147	06	06	207	06	06
028	02	2	088	43	RCL	148	01	1	208	01	1
029	01	1	089	27	27	149	05	5	209	03	3
030	69	OP	090	45	Y×	150	02	2	210	00	0
031	04	04	091	03	3	151	04	4	211	04	4
032	43	RCL	092	65	×	152	69	OP	212	69	OP
033	28	28	093	93	.	153	04	04	213	04	04
034	33	X²	094	02	2	154	43	RCL	214	43	RCL
035	55	÷	095	06	6	155	56	56	215	55	55
036	53	(	096	09	9	156	69	OP	216	69	OP
037	94	+/-	097	94	+/-	157	06	06	217	06	06
038	85	+	098	95	=	158	01	1	218	01	1
039	01	1	099	42	STO	159	03	3	219	05	5
040	54	)	100	27	27	160	00	0	220	02	2
041	65	×	101	43	RCL	161	02	2	221	04	4
042	53	(	102	26	26	162	69	OP	222	69	OP
043	43	RCL	103	49	PRD	163	04	04	223	04	04
044	00	00	104	27	27	164	43	RCL	224	43	RCL
045	75	-	105	55	÷	165	53	53	225	27	27
046	04	4	106	43	RCL	166	69	OP	226	65	×
047	54	)	107	00	00	167	06	06	227	43	RCL
048	55	÷	108	65	×	168	01	1	228	25	25
049	03	3	109	53	(	169	05	5	229	34	ГX
050	95	=	110	43	RCL	170	02	2	230	95	=
051	69	OP	111	00	00	171	04	4	231	69	OP
052	06	06	112	85	+	172	69	OP	232	06	06
053	98	ADV	113	43	RCL	173	04	04	233	98	ADV
054	01	1	114	20	20	174	43	RCL	234	03	3
055	75	-	115	65	×	175	27	27	235	05	5
056	43	RCL	116	43	RCL	176	65	×	236	69	OP
057	29	29	117	30	30	177	43	RCL	237	04	04
058	35	1/X	118	33	X²	178	20	20	238	69	OP
059	95	=	119	85	+	179	34	ГX	239	05	05

TABLE 4-39

Program 434 Listing

240	00	0	300	98	ADV	360	45	45
241	02	2	301	98	ADV	361	85	+
242	00	0	302	98	ADV	362	43	RCL
243	03	3	303	91	R/S	363	55	55
244	69	OP	304	76	LBL	364	65	×
245	04	04	305	12	B	365	43	RCL
246	43	RCL	306	04	4	366	46	46
247	21	21	307	04	4	367	95	=
248	55	÷	308	00	0	368	42	STO
249	53	(	309	02	2	369	57	57
250	43	RCL	310	69	OP	370	58	FIX
251	20	20	311	04	04	371	02	02
252	65	×	312	91	R/S	372	69	OP
253	43	RCL	313	42	STO	373	06	06
254	23	23	314	44	44	374	04	4
255	54	)	315	69	OP	375	05	5
256	34	ГX	316	06	06	376	69	OP
257	95	=	317	04	4	377	04	04
258	69	OP	318	04	4	378	91	R/S
259	06	06	319	00	0	379	42	STO
260	00	0	320	03	3	380	47	47
261	02	2	321	69	OP	381	69	OP
262	00	0	322	04	04	382	06	06
263	04	4	323	91	R/S	383	07	7
264	69	OP	324	68	NOP	384	05	5
265	04	04	325	68	NOP	385	69	OP
266	43	RCL	326	42	STO	386	04	04
267	22	22	327	45	45	387	43	RCL
268	55	÷	328	69	OP	388	47	47
269	53	(	329	06	06	389	75	-
270	43	RCL	330	04	4	390	43	RCL
271	20	20	331	04	4	391	57	57
272	65	×	332	00	0	392	95	=
273	43	RCL	333	04	4	393	69	OP
274	25	25	334	69	OP	394	06	06
275	54	)	335	04	04	395	98	ADV
276	34	ГX	336	91	R/S	396	22	INV
277	95	=	337	42	STO	397	58	FIX
278	69	OP	338	46	46	398	61	GTO
279	06	06	339	69	OP	399	12	B
280	00	0	340	06	06			
281	03	3	341	04	4			
282	00	0	342	05	5			
283	04	4	343	01	1			
284	69	OP	344	05	5			
285	04	04	345	69	OP			
286	43	RCL	346	04	04			
287	24	24	347	43	RCL			
288	55	÷	348	52	52			
289	53	(	349	85	+			
290	43	RCL	350	43	RCL			
291	23	23	351	53	53			
292	65	×	352	65	×			
293	43	RCL	353	43	RCL			
294	25	25	354	44	44			
295	54	)	355	85	+			
296	34	ГX	356	43	RCL			
297	95	=	357	54	54			
298	69	OP	358	65	×			
299	06	06	359	43	RCL			

TABLE 4-40

Program 435 Listing

000	76	LBL	060	05	05	120	13	13	180	22	INV
001	11	A	061	43	RCL	121	99	PRT	181	44	SUM
002	01	1	062	41	41	122	91	R/S	182	33	33
003	04	4	063	42	STO	123	76	LBL	183	43	RCL
004	99	PRT	064	06	06	124	16	A'	184	46	46
005	43	RCL	065	69	OP	125	91	R/S	185	65	×
006	42	42	066	13	13	126	42	STO	186	43	RCL
007	42	STO	067	99	PRT	127	44	44	187	45	45
008	01	01	068	01	1	128	69	OP	188	95	=
009	43	RCL	069	03	3	129	30	30	189	22	INV
010	43	43	070	99	PRT	130	43	RCL	190	44	SUM
011	42	STO	071	43	RCL	131	00	00	191	37	37
012	02	02	072	30	30	132	99	PRT	192	91	R/S
013	43	RCL	073	42	STO	133	43	RCL	193	99	PRT
014	00	00	074	01	01	134	44	44	194	42	STO
015	42	STO	075	43	RCL	135	99	PRT	195	47	47
016	03	03	076	31	31	136	22	INV	196	22	INV
017	43	RCL	077	42	STO	137	44	SUM	197	44	SUM
018	30	30	078	02	02	138	30	30	198	42	42
019	42	STO	079	43	RCL	139	33	X^2	199	33	X^2
020	04	04	080	33	33	140	22	INV	200	22	INV
021	43	RCL	081	42	STO	141	44	SUM	201	44	SUM
022	31	31	082	06	06	142	31	31	202	43	43
023	42	STO	083	69	OP	143	91	R/S	203	43	RCL
024	05	05	084	13	13	144	99	PRT	204	47	47
025	43	RCL	085	99	PRT	145	42	STO	205	65	×
026	34	34	086	02	2	146	45	45	206	43	RCL
027	42	STO	087	03	3	147	22	INV	207	44	44
028	06	06	088	99	PRT	148	44	SUM	208	95	=
029	69	OP	089	43	RCL	149	35	35	209	22	INV
030	13	13	090	35	35	150	33	X^2	210	44	SUM
031	99	PRT	091	42	STO	151	22	INV	211	34	34
032	02	2	092	01	01	152	44	SUM	212	43	RCL
033	04	4	093	43	RCL	153	36	36	213	47	47
034	99	PRT	094	36	36	154	43	RCL	214	65	×
035	43	RCL	095	42	STO	155	45	45	215	43	RCL
036	35	35	096	02	02	156	65	×	216	45	45
037	42	STO	097	43	RCL	157	43	RCL	217	95	=
038	04	04	098	37	37	158	44	44	218	22	INV
039	43	RCL	099	42	STO	159	95	=	219	44	SUM
040	36	36	100	06	06	160	22	INV	220	38	38
041	42	STO	101	69	OP	161	44	SUM	221	43	RCL
042	05	05	102	13	13	162	32	32	222	47	47
043	43	RCL	103	99	PRT	163	91	R/S	223	65	×
044	38	38	104	01	1	164	99	PRT	224	43	RCL
045	42	STO	105	02	2	165	42	STO	225	46	46
046	06	06	106	99	PRT	166	46	46	226	95	=
047	69	OP	107	43	RCL	167	22	INV	227	22	INV
048	13	13	108	30	30	168	44	SUM	228	44	SUM
049	99	PRT	109	42	STO	169	39	39	229	41	41
050	03	3	110	04	04	170	33	X^2	230	98	ADV
051	04	4	111	43	RCL	171	22	INV	231	61	GTO
052	99	PRT	112	31	31	172	44	SUM	232	16	A'
053	43	RCL	113	42	STO	173	40	40	233	00	0
054	39	39	114	05	05	174	43	RCL	234	00	0
055	42	STO	115	43	RCL	175	46	46			
056	04	04	116	32	32	176	65	×			
057	43	RCL	117	42	STO	177	43	RCL			
058	40	40	118	06	06	178	44	44			
059	42	STO	119	69	OP	179	95	=			

TABLE 4-41

Program 441 Listing

000	76	LBL	060	91	R/S	120	43	RCL	180	47	47
001	11	A	061	42	STO	121	62	62	181	55	÷
002	91	R/S	062	61	61	122	95	=	182	43	RCL
003	42	STO	063	44	SUM	123	44	SUM	183	00	00
004	58	58	064	50	50	124	56	56	184	95	=
005	69	OP	065	33	X²	125	43	RCL	185	42	STO
006	20	20	066	44	SUM	126	61	61	186	10	10
007	43	RCL	067	51	51	127	65	×	187	42	STO
008	00	00	068	43	RCL	128	43	RCL	188	16	16
009	99	PRT	069	58	58	129	62	62	189	43	RCL
010	43	RCL	070	65	×	130	99	PRT	190	42	42
011	58	58	071	43	RCL	131	95	=	191	75	-
012	99	PRT	072	61	61	132	44	SUM	192	43	RCL
013	44	SUM	073	95	=	133	57	57	193	38	38
014	38	38	074	44	SUM	134	98	ADV	194	65	×
015	33	X²	075	42	42	135	61	GTO	195	43	RCL
016	44	SUM	076	43	RCL	136	11	A	196	50	50
017	39	39	077	59	59	137	76	LBL	197	55	÷
018	91	R/S	078	65	×	138	12	B	198	43	RCL
019	42	STO	079	43	RCL	139	04	4	199	00	00
020	59	59	080	61	61	140	42	STO	200	95	=
021	99	PRT	081	95	=	141	07	07	201	42	STO
022	44	SUM	082	44	SUM	142	08	8	202	11	11
023	43	43	083	46	46	143	42	STO	203	42	STO
024	33	X²	084	43	RCL	144	01	01	204	20	20
025	44	SUM	085	60	60	145	43	RCL	205	43	RCL
026	44	44	086	65	×	146	39	39	206	44	44
027	43	RCL	087	43	RCL	147	75	-	207	75	-
028	58	58	088	61	61	148	43	RCL	208	43	RCL
029	65	×	089	99	PRT	149	38	38	209	43	43
030	43	RCL	090	95	=	150	33	X²	210	33	X²
031	59	59	091	44	SUM	151	55	÷	211	55	÷
032	95	=	092	49	49	152	43	RCL	212	43	RCL
033	44	SUM	093	91	R/S	153	00	00	213	00	00
034	40	40	094	42	STO	154	95	=	214	95	=
035	91	R/S	095	62	62	155	42	STO	215	42	STO
036	42	STO	096	44	SUM	156	08	08	216	13	13
037	60	60	097	52	52	157	43	RCL	217	43	RCL
038	44	SUM	098	33	X²	158	40	40	218	45	45
039	47	47	099	44	SUM	159	75	-	219	75	-
040	33	X²	100	53	53	160	43	RCL	220	43	RCL
041	44	SUM	101	43	RCL	161	38	38	221	43	43
042	48	48	102	58	58	162	65	×	222	65	×
043	43	RCL	103	65	×	163	43	RCL	223	43	RCL
044	58	58	104	43	RCL	164	43	43	224	47	47
045	65	×	105	62	62	165	55	÷	225	55	÷
046	43	RCL	106	95	=	166	43	RCL	226	43	RCL
047	60	60	107	44	SUM	167	00	00	227	00	00
048	95	=	108	54	54	168	95	=	228	95	=
049	44	SUM	109	43	RCL	169	42	STO	229	42	STO
050	41	41	110	59	59	170	09	09	230	14	14
051	43	RCL	111	65	×	171	42	STO	231	42	STO
052	59	59	112	43	RCL	172	12	12	232	17	17
053	65	×	113	62	62	173	43	RCL	233	43	RCL
054	43	RCL	114	95	=	174	41	41	234	48	48
055	60	60	115	44	SUM	175	75	-	235	75	-
056	99	PRT	116	55	55	176	43	RCL	236	43	RCL
057	95	=	117	43	RCL	177	38	38	237	47	47
058	44	SUM	118	60	60	178	65	×	238	33	X²
059	45	45	119	65	×	179	43	RCL	239	55	÷

TABLE 4-42

Program 442 Listing

240	43	RCL	300	95	=
241	00	00	301	42	STO
242	95	=	302	63	63
243	42	STO	303	43	RCL
244	18	18	304	55	55
245	43	RCL	305	75	-
246	49	49	306	43	RCL
247	75	-	307	43	43
248	43	RCL	308	65	×
249	47	47	309	43	RCL
250	65	×	310	52	52
251	43	RCL	311	55	÷
252	50	50	312	43	RCL
253	55	÷	313	00	00
254	43	RCL	314	95	=
255	00	00	315	42	STO
256	95	=	316	64	64
257	42	STO	317	91	R/S
258	19	19	318	00	0
259	42	STO	319	00	0
260	22	22			
261	43	RCL			
262	46	46			
263	75	-			
264	43	RCL			
265	43	43			
266	65	×			
267	43	RCL			
268	50	50			
269	55	÷			
270	43	RCL			
271	00	00			
272	95	=			
273	42	STO			
274	15	15			
275	42	STO			
276	21	21			
277	43	RCL			
278	51	51			
279	75	-			
280	43	RCL			
281	50	50			
282	33	X^2			
283	55	÷			
284	43	RCL			
285	00	00			
286	95	=			
287	42	STO			
288	23	23			
289	43	RCL			
290	54	54			
291	75	-			
292	43	RCL			
293	38	38			
294	65	×			
295	43	RCL			
296	52	52			
297	55	÷			
298	43	RCL			
299	00	00			

TABLE 4-43

Program 443 Listing

000	61	GTO	060	36	PGM	120	91	R/S	180	99	PRT
001	12	B	061	02	02	121	81	RST	181	43	RCL
002	76	LBL	062	91	R/S	122	76	LBL	182	30	30
003	11	A	063	42	STO	123	12	B	183	65	×
004	43	RCL	064	29	29	124	42	STO	184	43	RCL
005	56	56	065	36	PGM	125	37	37	185	63	63
006	75	-	066	02	02	126	43	RCL	186	85	+
007	43	RCL	067	91	R/S	127	28	28	187	43	RCL
008	47	47	068	42	STO	128	65	×	188	33	33
009	65	×	069	30	30	129	43	RCL	189	65	×
010	43	RCL	070	36	PGM	130	63	63	190	43	RCL
011	52	52	071	02	02	131	85	+	191	64	64
012	55	÷	072	91	R/S	132	43	RCL	192	85	+
013	43	RCL	073	42	STO	133	29	29	193	43	RCL
014	00	00	074	31	31	134	65	×	194	35	35
015	95	=	075	36	PGM	135	43	RCL	195	65	×
016	42	STO	076	02	02	136	64	64	196	43	RCL
017	65	65	077	91	R/S	137	85	+	197	65	65
018	43	RCL	078	36	PGM	138	43	RCL	198	85	+
019	57	57	079	02	02	139	30	30	199	43	RCL
020	75	-	080	91	R/S	140	65	×	200	36	36
021	43	RCL	081	42	STO	141	43	RCL	201	65	×
022	50	50	082	32	32	142	65	65	202	43	RCL
023	65	×	083	36	PGM	143	85	+	203	66	66
024	43	RCL	084	02	02	144	43	RCL	204	95	=
025	52	52	085	91	R/S	145	31	31	205	42	STO
026	55	÷	086	42	STO	146	65	×	206	71	71
027	43	RCL	087	33	33	147	43	RCL	207	99	PRT
028	00	00	088	36	PGM	148	66	66	208	43	RCL
029	95	=	089	02	02	149	95	=	209	31	31
030	42	STO	090	91	R/S	150	42	STO	210	65	×
031	66	66	091	42	STO	151	69	69	211	43	RCL
032	43	RCL	092	34	34	152	98	ADV	212	63	63
033	53	53	093	36	PGM	153	99	PRT	213	85	+
034	75	-	094	02	02	154	43	RCL	214	43	RCL
035	43	RCL	095	91	R/S	155	29	29	215	34	34
036	52	52	096	36	PGM	156	65	×	216	65	×
037	33	X²	097	02	02	157	43	RCL	217	43	RCL
038	55	÷	098	91	R/S	158	63	63	218	64	64
039	43	RCL	099	36	PGM	159	85	+	219	85	+
040	00	00	100	02	02	160	43	RCL	220	43	RCL
041	95	=	101	91	R/S	161	32	32	221	36	36
042	42	STO	102	42	STO	162	65	×	222	65	×
043	67	67	103	35	35	163	43	RCL	223	43	RCL
044	36	PGM	104	36	PGM	164	64	64	224	65	65
045	02	02	105	02	02	165	85	+	225	85	+
046	13	C	106	91	R/S	166	43	RCL	226	43	RCL
047	25	CLR	107	42	STO	167	33	33	227	37	37
048	36	PGM	108	36	36	168	65	×	228	65	×
049	02	02	109	36	PGM	169	43	RCL	229	43	RCL
050	17	B'	110	02	02	170	65	65	230	66	66
051	01	1	111	91	R/S	171	85	+	231	95	=
052	36	PGM	112	36	PGM	172	43	RCL	232	42	STO
053	02	02	113	02	02	173	34	34	233	72	72
054	18	C'	114	91	R/S	174	65	×	234	99	PRT
055	36	PGM	115	36	PGM	175	43	RCL	235	91	R/S
056	02	02	116	02	02	176	66	66	236	00	0
057	91	R/S	117	91	R/S	177	95	=	237	00	0
058	42	STO	118	36	PGM	178	42	STO	238	00	0
059	28	28	119	02	02	179	70	70	239	00	0

215

TABLE 4-44

Program 444 Listing

000	76	LBL	060	73	73	120	43	RCL	180	75	75
001	11	A	061	35	1/X	121	63	63	181	33	X²
002	43	RCL	062	94	+/-	122	75	-	182	55	÷
003	52	52	063	85	+	123	43	RCL	183	53	(
004	55	÷	064	01	1	124	70	70	184	94	+/-
005	43	RCL	065	95	=	125	65	×	185	85	+
006	00	00	066	35	1/X	126	43	RCL	186	01	1
007	75	-	067	42	STO	127	64	64	187	54	)
008	43	RCL	068	76	76	128	75	-	188	65	×
009	69	69	069	65	×	129	43	RCL	189	43	RCL
010	65	×	070	01	1	130	71	71	190	73	73
011	43	RCL	071	93	.	131	65	×	191	55	÷
012	38	38	072	02	2	132	43	RCL	192	04	4
013	55	÷	073	01	1	133	65	65	193	95	=
014	43	RCL	074	03	3	134	75	-	194	69	OP
015	00	00	075	04	4	135	43	RCL	195	06	06
016	75	-	076	85	+	136	72	72	196	98	ADV
017	43	RCL	077	93	.	137	65	×	197	98	ADV
018	70	70	078	00	0	138	43	RCL	198	91	R/S
019	65	×	079	00	0	139	66	66	199	00	0
020	43	RCL	080	08	8	140	54	)	200	00	0
021	43	43	081	06	6	141	55	÷	201	00	0
022	55	÷	082	85	+	142	43	RCL			
023	43	RCL	083	43	RCL	143	73	73			
024	00	00	084	76	76	144	95	=			
025	75	-	085	33	X²	145	34	√X̄			
026	43	RCL	086	65	×	146	42	STO			
027	71	71	087	01	1	147	74	74			
028	65	×	088	93	.	148	49	PRD			
029	43	RCL	089	00	0	149	76	76			
030	47	47	090	00	0	150	33	X²			
031	55	÷	091	05	5	151	65	×			
032	43	RCL	092	06	6	152	43	RCL			
033	00	00	093	85	+	153	73	73			
034	75	-	094	43	RCL	154	55	÷			
035	43	RCL	095	76	76	155	43	RCL			
036	72	72	096	45	Y×	156	67	67			
037	65	×	097	03	3	157	95	=			
038	43	RCL	098	65	×	158	94	+/-			
039	50	50	099	93	.	159	85	+			
040	55	÷	100	02	2	160	01	1			
041	43	RCL	101	06	6	161	95	=			
042	00	00	102	09	9	162	34	√X̄			
043	95	=	103	94	+/-	163	42	STO			
044	42	STO	104	95	=	164	75	.75			
045	68	68	105	42	STO	165	69	OP			
046	98	ADV	106	76	76	166	06	06			
047	99	PRT	107	42	STO	167	03	3			
048	03	3	108	79	79	168	06	6			
049	01	1	109	03	3	169	69	OP			
050	69	OP	110	05	5	170	04	04			
051	04	04	111	69	OP	171	43	RCL			
052	43	RCL	112	04	04	172	74	74			
053	00	00	113	53	(	173	69	OP			
054	69	OP	114	43	RCL	174	06	06			
055	06	06	115	67	67	175	02	2			
056	75	-	116	75	-	176	01	1			
057	05	5	117	43	RCL	177	69	OP			
058	95	=	118	69	69	178	04	04			
059	42	STO	119	65	×	179	43	RCL			

TABLE 4-45
Program 445 Listing

000	76	LBL	060	29	29	120	01	1	180	01	1
001	11	A	061	65	×	121	03	3	181	05	5
002	01	1	062	43	RCL	122	00	0	182	02	2
003	03	3	063	38	38	123	02	2	183	04	4
004	00	0	064	65	×	124	69	OP	184	69	OP
005	01	1	065	43	RCL	125	04	04	185	04	04
006	69	OP	066	43	43	126	43	RCL	186	43	RCL
007	04	04	067	85	+	127	69	69	187	76	76
008	43	RCL	068	43	RCL	128	69	OP	188	65	×
009	68	68	069	30	30	129	06	06	189	43	RCL
010	69	OP	070	65	×	130	01	1	190	35	35
011	06	06	071	43	RCL	131	05	5	191	34	ΓX
012	01	1	072	38	38	132	02	2	192	95	=
013	05	5	073	65	×	133	04	4	193	69	OP
014	02	2	074	43	RCL	134	69	OP	194	06	06
015	04	4	075	47	47	135	04	04	195	01	1
016	69	OP	076	85	+	136	43	RCL	196	03	3
017	04	04	077	43	RCL	137	76	76	197	00	0
018	43	RCL	078	31	31	138	65	×	198	05	5
019	76	76	079	65	×	139	43	RCL	199	69	OP
020	55	÷	080	43	RCL	140	28	28	200	04	04
021	43	RCL	081	38	38	141	34	ΓX	201	43	RCL
022	00	00	082	65	×	142	95	=	202	72	72
023	65	×	083	43	RCL	143	69	OP	203	69	OP
024	53	(	084	50	50	144	06	06	204	06	06
025	43	RCL	085	85	+	145	01	1	205	01	1
026	00	00	086	43	RCL	146	03	3	206	05	5
027	85	+	087	33	33	147	00	0	207	02	2
028	43	RCL	088	65	×	148	03	3	208	04	4
029	28	28	089	43	RCL	149	69	OP	209	69	OP
030	65	×	090	43	43	150	04	04	210	04	04
031	43	RCL	091	65	×	151	43	RCL	211	43	RCL
032	38	38	092	43	RCL	152	70	70	212	76	76
033	33	X²	093	47	47	153	69	OP	213	65	×
034	85	+	094	85	+	154	06	06	214	43	RCL
035	43	RCL	095	43	RCL	155	01	1	215	37	37
036	32	32	096	34	34	156	05	5	216	34	ΓX
037	65	×	097	65	×	157	02	2	217	95	=
038	43	RCL	098	43	RCL	158	04	4	218	69	OP
039	43	43	099	43	43	159	69	OP	219	06	06
040	33	X²	100	65	×	160	04	04	220	98	ADV
041	85	+	101	43	RCL	161	43	RCL	221	98	ADV
042	43	RCL	102	50	50	162	76	76	222	98	ADV
043	35	35	103	85	+	163	65	×	223	91	R/S
044	65	×	104	43	RCL	164	43	RCL	224	00	0
045	43	RCL	105	36	36	165	32	32	225	00	0
046	47	47	106	65	×	166	34	ΓX	226	00	0
047	33	X²	107	43	RCL	167	95	=	227	00	0
048	85	+	108	47	47	168	69	OP			
049	43	RCL	109	65	×	169	06	06			
050	37	37	110	43	RCL	170	01	1			
051	65	×	111	50	50	171	03	3			
052	43	RCL	112	54	)	172	00	0			
053	50	50	113	54	)	173	04	4			
054	33	X²	114	34	ΓX	174	69	OP			
055	85	+	115	95	=	175	04	04			
056	02	2	116	42	STO	176	43	RCL			
057	65	×	117	77	77	177	71	71			
058	53	(	118	69	OP	178	69	OP			
059	43	RCL	119	06	06	179	06	06			

TABLE 4-46

Program 446 Listing

000	76	LBL	060	48	48	120	06	06	180	99	PRT
001	11	A	061	42	STO	121	69	OP	181	43	RCL
002	69	OP	062	05	05	122	13	13	182	38	38
003	00	00	063	43	RCL	123	99	PRT	183	42	STO
004	01	1	064	56	56	124	03	3	184	01	01
005	05	5	065	42	STO	125	04	4	185	43	RCL
006	99	PRT	066	06	06	126	99	PRT	186	39	39
007	43	RCL	067	69	OP	127	43	RCL	187	42	STO
008	52	52	068	13	13	128	47	47	188	02	02
009	42	STO	069	99	PRT	129	42	STO	189	43	RCL
010	01	01	070	04	4	130	01	01	190	40	40
011	43	RCL	071	05	5	131	43	RCL	191	42	STO
012	53	53	072	99	PRT	132	48	48	192	06	06
013	42	STO	073	43	RCL	133	42	STO	193	69	OP
014	02	02	074	50	50	134	02	02	194	13	13
015	43	RCL	075	42	STO	135	43	RCL	195	99	PRT
016	00	00	076	04	04	136	49	49	196	91	R/S
017	42	STO	077	43	RCL	137	42	STO	197	00	0
018	03	03	078	51	51	138	06	06	198	00	0
019	43	RCL	079	42	STO	139	69	OP	199	00	0
020	38	38	080	05	05	140	13	13	200	00	0
021	42	STO	081	43	RCL	141	99	PRT	201	00	0
022	04	04	082	57	57	142	01	1	202	00	0
023	43	RCL	083	42	STO	143	03	3	203	00	0
024	39	39	084	06	06	144	99	PRT	204	00	0
025	42	STO	085	69	OP	145	43	RCL	205	00	0
026	05	05	086	13	13	146	38	38	206	00	0
027	43	RCL	087	99	PRT	147	42	STO	207	00	0
028	54	54	088	01	1	148	04	04	208	00	0
029	42	STO	089	04	4	149	43	RCL	209	00	0
030	06	06	090	99	PRT	150	39	39	210	00	0
031	69	OP	091	43	RCL	151	42	STO	211	00	0
032	13	13	092	38	38	152	05	05			
033	99	PRT	093	42	STO	153	43	RCL			
034	02	2	094	01	01	154	41	41			
035	05	5	095	43	RCL	155	42	STO			
036	99	PRT	096	39	39	156	06	06			
037	43	RCL	097	42	STO	157	69	OP			
038	43	43	098	02	02	158	13	13			
039	42	STO	099	43	RCL	159	99	PRT			
040	04	04	100	42	42	160	02	2			
041	43	RCL	101	42	STO	161	03	3			
042	44	44	102	06	06	162	99	PRT			
043	42	STO	103	69	OP	163	43	RCL			
044	05	05	104	13	13	164	43	43			
045	43	RCL	105	99	PRT	165	42	STO			
046	55	55	106	02	2	166	04	04			
047	42	STO	107	04	4	167	43	RCL			
048	06	06	108	99	PRT	168	44	44			
049	69	OP	109	43	RCL	169	42	STO			
050	13	13	110	43	43	170	05	05			
051	99	PRT	111	42	STO	171	43	RCL			
052	03	3	112	01	01	172	45	45			
053	05	5	113	43	RCL	173	42	STO			
054	99	PRT	114	44	44	174	06	06			
055	43	RCL	115	42	STO	175	69	OP			
056	47	47	116	02	02	176	13	13			
057	42	STO	117	43	RCL	177	99	PRT			
058	04	04	118	46	46	178	01	1			
059	43	RCL	119	42	STO	179	02	2			

TABLE 4-47

Program 447 Listing

000	76	LBL	060	59	59	120	43	RCL	180	04	04
001	11	A	061	65	×	121	62	62	181	43	RCL
002	91	R/S	062	43	RCL	122	95	=	182	68	68
003	42	STO	063	60	60	123	22	INV	183	85	+
004	58	58	064	99	PRT	124	44	SUM	184	43	RCL
005	69	OP	065	95	=	125	54	54	185	69	69
006	30	30	066	22	INV	126	43	RCL	186	65	×
007	43	RCL	067	44	SUM	127	59	59	187	43	RCL
008	00	00	068	45	45	128	65	×	188	58	58
009	99	PRT	069	91	R/S	129	43	RCL	189	85	+
010	43	RCL	070	42	STO	130	62	62	190	43	RCL
011	58	58	071	61	61	131	95	=	191	70	70
012	99	PRT	072	22	INV	132	22	INV	192	65	×
013	22	INV	073	44	SUM	133	44	SUM	193	43	RCL
014	44	SUM	074	50	50	134	55	55	194	59	59
015	38	38	075	33	X²	135	43	RCL	195	85	+
016	33	X²	076	22	INV	136	60	60	196	43	RCL
017	22	INV	077	44	SUM	137	65	×	197	71	71
018	44	SUM	078	51	51	138	43	RCL	198	65	×
019	39	39	079	43	RCL	139	62	62	199	43	RCL
020	91	R/S	080	58	58	140	95	=	200	60	60
021	42	STO	081	65	×	141	22	INV	201	85	+
022	59	59	082	43	RCL	142	44	SUM	202	43	RCL
023	99	PRT	083	61	61	143	56	56	203	72	72
024	22	INV	084	95	=	144	43	RCL	204	65	×
025	44	SUM	085	22	INV	145	61	61	205	43	RCL
026	43	43	086	44	SUM	146	65	×	206	61	61
027	33	X²	087	42	42	147	43	RCL	207	95	=
028	22	INV	088	43	RCL	148	62	62	208	42	STO
029	44	SUM	089	59	59	149	99	PRT	209	78	78
030	44	44	090	65	×	150	95	=	210	69	OP
031	43	RCL	091	43	RCL	151	22	INV	211	06	06
032	58	58	092	61	61	152	44	SUM	212	04	4
033	65	×	093	95	=	153	57	57	213	05	5
034	43	RCL	094	22	INV	154	98	ADV	214	69	OP
035	59	59	095	44	SUM	155	61	GTO	215	04	04
036	95	=	096	46	46	156	11	A	216	91	R/S
037	22	INV	097	43	RCL	157	76	LBL	217	42	STO
038	44	SUM	098	60	60	158	12	B	218	62	62
039	40	40	099	65	×	159	91	R/S	219	69	OP
040	91	R/S	100	43	RCL	160	42	STO	220	06	06
041	42	STO	101	61	61	161	58	58	221	07	7
042	60	60	102	99	PRT	162	99	PRT	222	05	5
043	22	INV	103	95	=	163	91	R/S	223	69	OP
044	44	SUM	104	22	INV	164	42	STO	224	04	04
045	47	47	105	44	SUM	165	59	59	225	43	RCL
046	33	X²	106	49	49	166	99	PRT	226	62	62
047	22	INV	107	91	R/S	167	91	R/S	227	75	-
048	44	SUM	108	42	STO	168	42	STO	228	43	RCL
049	48	48	109	62	62	169	60	60	229	78	78
050	43	RCL	110	22	INV	170	99	PRT	230	95	=
051	58	58	111	44	SUM	171	91	R/S	231	69	OP
052	65	×	112	52	52	172	42	STO	232	06	06
053	43	RCL	113	33	X²	173	61	61	233	98	ADV
054	60	60	114	22	INV	174	99	PRT	234	61	GTO
055	95	=	115	44	SUM	175	04	4	235	12	B
056	22	INV	116	53	53	176	05	5	236	00	0
057	44	SUM	117	43	RCL	177	01	1	237	00	0
058	41	41	118	58	58	178	05	5	238	00	0
059	43	RCL	119	65	×	179	69	OP			

program are omitted here and data are placed in the proper registers by this program.

(4) PGM 02 C calculates the determinant which is automatically printed.

(5) CLR followed by PGM 02 B' calculates the inverse matrix. 1 PGM 02 C' prepares for display of the elements of the matrix. The order in which these elements appear is arbitrary and depends on "pivoting" during their calculation. For this reason, it is necessary to rely on the Library Module to select the elements in their proper sequence. The sequence PGM 02 R/S recalls each element in turn.

(6) The number of degrees of freedom is $n - 3$ for a three-variable equation. The bilinear equation has a fourth variable β (or log β) and $n - 4$ is correct.

(7) The bilinear equation requires three variables here.

(8) The sequence 401–447 calculates t. See example of Section III.E.

(9) Label C positions program operation to calculate y_C and Δ.

3. Program 424

(1) In this short nomenclature 13 means x, y, 23 means $x_2 y$, and 12 means $x_1 x_2$.

(2) It is a simple matter to place the sums in the proper registers and allow the built-in calculator program to calculate R.

(3) Op 13 calculates the correlation coefficient.

4. Programs 425, 426

(1) Label A selects x_1 of the three-variable equation to become X of the two-variable equation.

(2) R_{30} stores the input for the label.

(3) Flag 1 is set to indicate that x_1 of the three-variable equation is x of the two-variable equation.

(4) Label B selects x_2 as a component of the two-variable equation. If flag 1 is set, x_2 becomes y; if flag 2 is not set, x_2 becomes x of the two-variable equation.

(5) Label PGM performs the calculations.

(6) The sequence 128–173 calculates t.

(7) Program operation moves to label π, where F is calculated for addition of the third variable.

(8) Label GE prints the results for the two-variable equation in $x_1 x_2$.

(9) Label C selects y as the second variable in the two-variable equation.

(10) Flag 2 is set to indicate that the second variable in the two-variable equation is y.

(11) If flag 1 is set, x_1 is the first variable in the two-variable equation. The second variable must be y since label C was pressed.

(12) Label IFF selects $\Sigma x_1 y$ to store in R_{06}.

(13) Label π completes the calculations.

(14) Label D calculates y_C and Δ for the two-variable equation.

5. Programs 427, 428

(1) The raw data consisting of X and y values for up to and including 14 points are stored in R_{40}–R_{67} by the pointer in R_{01}.

(2) Label A begins the calculation.

(3) The value of n is saved in R_{68}. Repartitioning to include only 30 memories followed by CMS clears the first 30 data storage registers. (The value of last F in R_{36} must be saved but the sums in R_{21}–R_{29} must be removed.) Repartitioning again to include 70 registers allows n to be recalled and stored for use in R_{00}. The pointer is set up again in R_{01}. The sequence 0 EXC 68 STO 00 places 0 in R_{68} and n in R_{00}.

(4) Label DMS begins calling the data. Label E' (step 392) places 0 in the t register. Data are called until R_{68} is reached, which contains 0 (step 038).

(5) The sequence 057–070 gives a value of log $(\beta P + 1)$ to x_2. Log P is X_1. Log β is stored in R_{69} or, alternatively, it is 0 on the first iteration.

(6) The sequence 073–114 sets up the sums in R_{21}–R_{29} just as in program 422.

(7) Label IFF performs the calculation as for the three-variable equation except that now there are four degrees of freedom (step 268). The value of log β is printed, followed by R and then the value of F.

(8) The sequence 312–390 uses flags 1, 2, and 3 and the INV $X \geq t$ keys to continue the iteration until a maximum value of F is found and passed by removing 0.01 from log β (see decision map, Fig. 4-4). Finally, 0.01 is added to log β and the final values of log β, R, and F are printed. When program operation stops, the data registers are all set up for a three-variable equation. Programs 421, 422, and 423 are now used to obtain the coefficients and their 95% confidence limits using four degrees of freedom.

(9) R_{17} is zero for the first iteration.

6. Programs for Four and Five Variables

Programs 43 and 44 are designed in the same manner as programs 421, 422, and 423 except that more memories and more steps are required for the increased number of variables. Table 4-24 lists the register contents and Section IV.E provides a summary of the label operations. The equations are provided in Section II.

REFERENCES

1. N. B. Chapman and J. Shorter, Eds. "Correlation Analysis in Chemistry." Plenum Press, New York, 1978.
2. Y. C. Martin, "Quantitative Drug Design." Marcel Dekker, New York, 1978.
3. N. R. Draper and H. Smith, "Applied Regression Analysis." Wiley, New York, 1966.
4. "Personal Programming," TI Programmable 58/59 Owner's Manual, Texas Instruments, Inc., Dallas, Texas, 1977.
5. O. L. Davies, "Statistical Methods in Research and Production." Hafner, New York, 1967.
6. E. A. Coats, S. L. Milstein, G. Holbein, J. McDonald, R. Reed, and H. G. Petering, *J. Med. Chem.* **19**, 131, 1976.
7. P. B. M. Timmermans, A. Brands, and P. A. van Zwieten, Naunyn-Schmiedeberg's Arch. Pharmacol. **300**, 217, 1977.
8. H. Kubinyi, Lipophilicity and drug activity. *In* "Progress in Drug Research" (E. Jucker, ed.), Vol. 23, p. 97. Birkhäuser, Basel, 1979.
9. T. R. Harshbarger, "Introductory Statistics: A Decision Map." Macmillan, New York, 1977.
10. C. Hansch, A. Leo, S. H. Unger, K. H. Kim, D. Nikaitani, and E. J. Lien, *J. Med. Chem.* **16**, 1207, 1973.

Index

Numbers in bold type refer to program numbers.